Texts in Theoretical Computer Science
An EATCS Series

H0246025

José Luis Balcázar
Josep Díaz
Joaquim Gabarró

Structural
Complexity I

Second, Revised Edition
With 60 Figures

 Springer

Authors

Prof. Dr. José Luis Balcázar
Prof. Dr. Josep Díaz
Prof. Dr. Joaquim Gabarró
Department of Software (LSI)
Universitat Politècnica de Catalunya
Pau Gargallo, 5, E-08028 Barcelona, Spain

Series Editors

Prof. Dr. Wilfried Brauer
Fakultät für Informatik, Technische Universität München
Arcisstrasse 21, D-80333 München, Germany

Prof. Dr. Grzegorz Rozenberg
Institute of Applied Mathematics and Computer Science
University of Leiden, Niels-Bohr-Weg 1, P.O. Box 9512
NL-2300 RA Leiden, The Netherlands

Prof. Dr. Arto Salomaa
The Academy of Finland
Department of Mathematics, University of Turku
FIN-20 500 Turku, Finland

Library of Congress Cataloging-in-Publication Data
Balcázar, José Luis. Structural complexity I/José Luis Balcázar, Josep Díaz, Joaquim
Gabarró. p. cm. – (Texts in theoretical computer science. An EATCS series)
Includes bibliographical references and index.
ISBN-13: 978-3-642-79237-3   e-ISBN-13: 978-3-642-79235-9
DOI: 10.1007/ 978-3-642-79235-9
1. Computer science. 2. Computational complexity. I. Díaz, J. (Josep), 1950-   . II. Gabarró,
Joaquim.   III. Title.   IV. Series. QA76..B257   1994 511.3—dc20 94-36688 CIP

© Springer-Verlag Berlin Heidelberg 1988, 1995
Softcover reprint of the hardcover 2nd edition 1995

SPIN: 10465773   45/3140 - 5 4 3 2 1 0 - Printed on acid-free paper

A Note from the Series Editors

The *EATCS Monographs on Theoretical Computer Science* series already has a fairly long tradition of more than thirty volumes over ten years. Many of the volumes have turned out to be useful also as textbooks. To give even more freedom for prospective authors and more choice for the audience, a Texts series has been branched off:

Texts in Theoretical Computer Science. An EATCS Series.

Texts published in this series are intended mostly for the graduate level. Typically, an undergraduate background in computer science will be assumed. However, the background required will vary from topic to topic, and some books will be self-contained. The texts will cover both modern and classical areas with an innovative approach that may give them additional value as monographs. Most books in this series will have examples and exercises.

The original series will continue as

Monographs in Theoretical Computer Science. An EATCS Series.

Books published in this series present original research or material of interest to the research community and graduate students. Each volume is normally a uniform monograph rather than a compendium of articles. The series also contains high-level presentations of special topics. Nevertheless, as research and teaching usually go hand in hand, these volumes may still be useful as textbooks, too.

The present volume is an excellent example of a textbook that, together with its second volume, has also considerable value as a monograph. Enjoy!

W. Brauer, G. Rozenberg, A. Salomaa

Preface to the Second Edition

In the six years since the first edition of this book was published, the field of Structural Complexity has grown quite a bit. However, we are keeping this volume at the same basic level that it had in the first edition, and the only new result incorporated as an appendix is the closure under complementation of nondeterministic space classes, which in the previous edition was posed as an open problem. This result was already included in our Volume II, but we feel that due to the basic nature of the result, it belongs to this volume. There are of course other important results obtained during these last six years. However, as they belong to new areas opened in the field they are outside the scope of this fundamental volume.

Other changes in this second edition are the update of some Bibliographical Remarks and references, correction of many mistakes and typos, and a renumbering of the definitions and results. Experience has shown us that this new numbering is a lot more friendly, and several readers have confirmed this opinion. For the sake of the reader of Volume II, where all references to Volume I follow the old numbering, we have included here a table indicating the new number corresponding to each of the old ones.

The authors are pleased by the fact that the book has been adopted as a textbook at several universities, and a lot of people have sent comments and reported mistakes, that have greatly improved this second edition. Unfortunately we cannot enumerate all of them here. However, we would like to express our gratitude to Ray Greenlaw, Torben Hagerup, Elvira Mayordomo, and Jan Van den Bussche for their very detailed reports, as well as to so many others who pointed out errors worthy of correction. Finally, we would like to thank Dr. H. Wössner and Mrs. I. Mayer of Springer-Verlag for their assistance.

Barcelona, November 1994

J.L. Balcázar
J. Díaz
J. Gabarró

Preface

Since the achievement of a formal definition of the concept of "algorithm", the Mathematical Theory of Computation has developed into a broad and rich discipline. The notion of "complexity of an algorithm" yields an important area of research, known as Complexity Theory, that can be approached from several points of view. Some of these are briefly discussed in the Introduction and, in particular, our view of the "Structural" approach is outlined there.

We feel the subject is mature enough to permit collecting and interrelating many of the results in book form. Let us point out that a substantial part of the knowledge in Structural Complexity Theory can be found only in specialized journals, symposia proceedings, and monographs like doctoral dissertations or similar texts, mostly unpublished. We believe that a task to be done soon is a systematization of the interconnections between all the research lines; this is a serious and long task. We hope that the two volumes of this book can serve as a starting point for this systematization process.

This book assumes as a prerequisite some knowledge of the basic models of computation, as taught in an undergraduate course on Automata Theory, Formal Language Theory, or Theory of Computation. Certainly, some mathematical maturity is required, and previous exposure to programming languages and programming techniques is desirable. Most of the material of Volume I can be successfully presented in a senior undergraduate course; Volumes I and II should be suitable for a first graduate course. Some sections lead to a point in which very little additional work suffices to be able to start research projects. In order to ease this step, an effort has been made to point out the main references for each of the results presented in the text.

Thus, each chapter ends with a section entitled "Bibliographical Remarks", in which the relevant references for the chapter are briefly commented upon. These sections might also be of interest to those wanting an overview of the evolution of the field. Additionally, each chapter (excluding the first two, which are intended to provide some necessary background) includes a section of exercises. The reader is encouraged to spend some time on them. Some results presented as exercises are used later in the text; however, this is the exception, not the rule. Many exercises are devoted to presenting a definition and some properties of an interesting concept. Usually, a reference is provided for the most interesting and for the most useful exercises. Some exercises are marked with a * to indicate that, to the best knowledge of the authors, the solution has a certain degree of difficulty.

This book originated from a somewhat incomplete set of lecture notes for a series of lectures given at the Institute of Computer Science and Cybernetics of Hanoi, Vietnam, during the summer of 1985. We would like to thank the head of the Institute at the time, Dr. Phan Dinh Dieu, for the invitation, as well as for the comments and remarks made during the lectures.

We have benefited from conversations with many friends and colleagues. In particular we would like to express our gratitude to Ron Book, Rafael Casas, Uwe Schöning, and many other colleagues for their suggestions and help, and to Peter van Emde Boas, Jacobo Torán, Carme Torras, the students of our Department, and an anonymous referee for pointing out many corrections which have improved the quality of the manuscript. We also would like to thank Rosa Martín and the other staff of the Laboratori de Càlcul de la Facultat d'Informàtica de Barcelona, for their logistic help in installing TEX on our computer system.

Barcelona, October 1987   *J.L. Balcázar*
                                                            J. Diaz
                                                            J. Gabarró

Contents

Cross References

Old Number → New Number (see Preface to the 2nd Edition)

Chapter 1

Definition			Example	Lemma
1.1 →1.1	1.13 →1.16	1.25 →1.37	1.1→1.11	1.1→1.10
1.2 →1.2	1.14 →1.17	1.26 →1.38	1.2→1.21	1.2→1.15
1.3 →1.3	1.15 →1.18	1.27 →1.39	1.3→1.26	1.3→1.29
1.4 →1.4	1.16 →1.20	1.28 →1.40	1.4→1.27	1.4→1.30
1.5 →1.5	1.17 →1.22	1.29 →1.41	1.5→1.46	*Theorem*
1.6 →1.6	1.18 →1.23	1.30 →1.42	1.6→1.48	1.1→1.35
1.7 →1.7	1.19 →1.24	1.31 →1.43	1.7→1.51	*Corollary*
1.8 →1.8	1.20 →1.25	1.32 →1.44	*Proposition*	1.1→1.36
1.9 →1.9	1.21 →1.28	1.33 →1.45	1.1→1.19	
1.10 →1.12	1.22 →1.32	1.34 →1.47	1.2→1.31	
1.11 →1.13	1.23 →1.33	1.35 →1.49		
1.12 →1.14	1.24 →1.34	1.36 →1.50		

Chapter 2

Definition	Proposition	Lemma	Theorem	
2.1→2.1	2.1→2.2	2.1→2.14	2.1→2.12	2.9 →2.27
2.2→2.6	2.2→2.3	2.2→2.19	2.2→2.13	2.10 →2.29
2.3→2.11	2.3→2.4	2.3→2.23	2.3→2.15	2.11 →2.30
2.4→2.18	2.4→2.7	2.4→2.25	2.4→2.16	*Corollary*
2.5→2.24	2.5→2.8		2.5→2.17	2.1 →2.28
Example	2.6→2.9		2.6→2.20	
2.1→2.5	2.7→2.10		2.7→2.21	
2.2→2.22			2.8→2.26	

Chapter 3

Definition	Example	Lemma	Theorem		Corollary
3.1→3.1	3.1→3.13	3.1→3.22	3.1→3.3	3.10 →3.30	3.1→3.8
3.2→3.5	*Proposition*	3.2→3.27	3.2→3.4	3.11 →3.32	3.2→3.19
3.3→3.6	3.1→3.2	3.3→3.28	3.3→3.9	3.12 →3.37	3.3→3.31
3.4→3.10	3.2→3.7	3.4→3.35	3.4→3.14	3.13 →3.38	3.4→3.33
3.5→3.15	3.3→3.11		3.5→3.21		3.5→3.36
3.6→3.17	3.4→3.12		3.6→3.23		3.6→3.39
3.7→3.20	3.5→3.16		3.7→3.24		
3.8→3.25	3.6→3.18		3.8→3.26		
3.9→3.34			3.9→3.29		

Chapter 4

Definition		Proposition	Theorem	Corollary
4.1→4.1	4.8 →4.15	4.1→4.3	4.1→4.16	4.1→4.6
4.2→4.4	4.9 →4.17	4.2→4.5	4.2→4.19	4.2→4.20
4.3→4.7	4.10 →4.23	4.3→4.11	4.3→4.21	4.3→4.22
4.4→4.8	4.11 →4.26	4.4→4.13	4.4→4.24	4.4→4.25
4.5→4.9	4.12 →4.28	4.5→4.14	4.5→4.29	
4.6→4.10	*Example*	4.6→4.18		
4.7→4.12	4.1 →4.2	4.7→4.27		

Chapter 5

Definition		Example	Proposition	Theorem	
5.1→5.1	5.9 →5.14	5.1→5.2	5.1→5.38	5.1→5.19	5.9 →5.31
5.2→5.5	5.10 →5.18	5.2→5.3	*Lemma*	5.2→5.21	5.10 →5.33
5.3→5.6	5.11 →5.22	5.3→5.4	5.1→5.35	5.3→5.24	5.11 →5.36
5.4→5.7	5.12 →5.23	5.4→5.10		5.4→5.25	5.12 →5.40
5.5→5.8	5.13 →5.30	5.5→5.12		5.5→5.26	*Corollary*
5.6→5.9	5.14 →5.32	5.6→5.15		5.6→5.27	5.1 →5.20
5.7→5.11	5.15 →5.37	5.7→5.16		5.7→5.28	5.2 →5.34
5.8→5.13	5.16 →5.39	5.8→5.17		5.8→5.29	

Chapter 6

Definition	Example	Proposition	Lemma	Theorem	Corollary
6.1→6.1	6.1→6.2	6.1→6.5	6.1→6.11	6.1→6.3	6.1→6.7
6.2→6.4		6.2→6.6	6.2→6.12	6.2→6.13	6.2→6.18
6.3→6.10		6.3→6.8		6.3→6.16	6.3→6.23
6.4→6.14		6.4→6.9		6.4→6.17	6.4→6.24
6.5→6.15		6.5→6.19		6.5→6.22	6.5→6.29
6.6→6.21		6.6→6.20		6.6→6.28	6.6→6.30
6.7→6.25		6.7→6.27			
6.8→6.26					

Chapter 7

Definition	Proposition	Lemma	Theorem	Corollary
7.1→7.1	7.1→7.3	7.1→7.5	7.1→7.4	7.1→7.11
7.2→7.2	7.2→7.8	7.2→7.6	7.2→7.9	
		7.3→7.7	7.3→7.10	

Chapter 8

Definition	Proposition	Lemma	Theorem	Corollary
8.1→8.1	8.1→8.3	8.1→8.7	8.1→8.5	8.1→8.11
8.2→8.2	8.2→8.4	8.2→8.16	8.2→8.8	8.2→8.12
8.3→8.18	8.3→8.6	8.3→8.20	8.3→8.9	8.3→8.15
8.4→8.19	8.4→8.14		8.4→8.10	
	8.5→8.17		8.5→8.13	
			8.6→8.21	

Introduction

In its most general sense, Complexity Theory encompasses several quite different approaches. All of them have in common two characteristics, which can be considered as a general definition of the field: first, they study algorithms and the problems solved by them; second, they focus on the computational resources required by these algorithms.

The word "algorithm" must be taken here in a very broad sense. Each of the approaches to Complexity Theory may have its own concepts of algorithm, and some of them are equivalent while others are not. Some branches of the field define algorithms as programs in some appropriate (usually high-level) programming language, or at least as pieces of text very close to these programs. For example, this is usually done to evaluate the efficiency of given data structures or of methodologies for the design of programs. Others use mathematical models, some of them close to real computers, others quite different. Each model is adequate to a different use, and it is natural to choose for each use the most convenient model. Hence each approach defines "algorithm" in one or more among several possible ways, looking for the one best suited to each of the problems studied.

In our work, the algorithms will range from abstract models of computing agents (more precisely Turing machines) used to define complexity classes, to high-level pseudo-code used to specify concrete algorithms. Models of boolean digital computation will also be used to define complexity classes.

Thus, this is what we mean here by "algorithm". Somewhat easier to describe is what we mean by "resources": in general, we mean magnitudes of measurable entities which make computation possible; in particular, and depending on the model of the algorithm, we mean magnitudes like computation time, memory space required, or number of elementary instructions to be performed (as in boolean circuits).

The approach we take has been called quite frequently, during the last few years, "Structural Complexity". We fully adopt the term, and we explain it below, by indicating the characteristics and the interest of this approach, as compared to other approaches. Let us briefly review some of these other approaches, classifying them (somewhat artificially) by their "abstraction level". More information about some of them is given in the discussion following the list.

1. Let us start from "concrete" complexity, which can be described as a systematized set of techniques for evaluating the resources, mainly space and time, required by given concrete algorithms. There are two main approaches to this study: the "worst case", i.e. the "most pessimistic" assumption about the behavior of the algorithms, and the "average case", i.e. the study of the expected behavior of a given algorithm under a given probability distribution of the instances of the problem. This approach is more difficult to follow.

2. Still within concrete complexity, but requiring a higher level of abstraction, is the question of finding lower and upper bounds for the resources required to solve some particular, fixed problems. If it is successful, this approach may allow us to prove some algorithm to be *optimal* for the problem it solves, by matching its resource requirements with the lower bound. Proving such a lower bound requires that one abstract from the particular algorithm being used, and that one find a characteristic which is common to *all* the algorithms solving the problem in no matter how clever a form. This area is also of great difficulty, involving complex combinatorial techniques, and is progressing slowly although steadily.

3. Of course, if abstract models of algorithms are to be used, then their pairwise relationships must be established, comparing their computational power in terms of the increase of resources required to simulate one model by another. This gives a broad area of research, in which more and more sophisticated simulations among different models are being designed.

4. The approach just mentioned gives rise to computational complexity, in which the main body of this work is framed. Here, we abstract from particular problems by considering them just as formal languages over some alphabet. This study is based on bounding particular resources like computation time and memory space.

5. Another closely related area is the study of the combinatorial properties of the problems and of their instances. The boolean (digital) models of computation are highly useful in this field, which is related both to computational complexity and to the area of concrete complexity and proofs of lower bounds, as shown in Chapter 5 of this volume.

6. At the more abstract level, we find the axiomatic approach of Blum, which abstracts even from the actual notion of "resource" by giving a machine-independent formulation, and obtains results of interest for all the other approaches.

7. Finally, in a different (but related) vein, there is an application of the concept of "complexity" no longer to formal languages, but to the words constituting them. This concept, which measures in some sense the "amount of information" contained in a given word, is called Kol-

mogorov complexity. The relationship between this concept and our approach will be studied in Volume II.

Measuring resources is a basic tool; however, it is not *our* tool. Measures provide "quantitative" results: such and such an algorithm, such and such a problem, such and such a model of computation requires at most, or at least, such an amount of resources. We want to transcend this sort of result by asking questions about the *inherent properties* of the problems to be solved, and how they relate to the quantitative aspects of the algorithms solving them. In other words, we want to understand the "qualitative" properties of the problems.

Our way of doing this is to look for more or less well known *mathematical structures* inside the problems and inside the classes of problems. Each of these mathematical structures will furnish us with some particular "universal" elements: tops or bottoms in partial orders, lattices, and semilattices; comparable and incomparable elements; sets as domains, ranges, or graphs of certain functions. We look for properties that can be characterized in as many different terms as possible, so that each characterization explains an unclear (and sometimes unexpected) relationship. In this way, we hope to deepen our understanding of the many processes arising in Complexity Theory which remain insufficiently explained.

It should be mentioned that many of the concepts which arise in Structural Complexity originate in Recursive Function Theory, where they helped to establish the main results and many correct intuitions about the notion of algorithm and its hidden implications. Let us develop briefly some thoughts about the similarities of this discipline with Complexity Theory before continuing our exposition about structural concepts.

Recursive Function Theory was motivated by the achievement of a formal notion of algorithm, and "recursive" (i.e. algorithmically solvable) problems were defined. After having this definition available, this theory started to study and classify the nonrecursive (unsolvable) problems. Interest was centered on the recursively invariant notions, and hence nontrivial solvable problems formed only one class. Therefore, Recursive Function Theory studied the nonrecursive recursively enumerable problems and also the hierarchies classifying problems of higher difficulty into the "degrees of unsolvability": arithmetic, hyper-arithmetic, analytic,... The study of these problems led to the invention of priority methods and other sophisticated techniques.

With the increase of the use of actual computers, and their great development, the notion of algorithm acquired much more importance, and, simultaneously, showed itself to be insufficient. The notion of "feasible algorithm" was intuitively needed as soon as real computations began to be performed and physical machines programmed. It was clear that recursive problems needing centuries of computer time for solving instances with small, simple data were too difficult to be considered solvable on realistic grounds. A classification of the recursive problems was needed.

After some initial work, like the definition and study of the Grzegorczyk classes, interest focused on the analysis of problems in terms of the resources needed by the algorithms solving them. This idea allowed the internal structure of the class of recursive problems to be investigated. In the sixties, a notion of "feasibility" was developed and gradually accepted, consisting of algorithms having computation time polynomial in the size of the input. In the early seventies, polynomial time reductions were defined, and the classes related to these notions began to be studied.

And in a curious analogy with Recursive Function Theory, once the notion of "feasible problems" had been defined, major efforts were spent on the problems not known to be feasible. The polynomial time hierarchy and other concepts analogous to those of Recursive Function Theory were defined in the polynomial setting, building a full Theory of Complexity—with an important difference: the simpler problems of Recursive Function Theory, when translated into Complexity Theory, became difficult enough to challenge for years the whole research community. Right now, we consider that new insights and a deeper understanding of the central problems and their implications are required to solve them.

We expect some of these new insights to arise from a careful study of the structural properties of the complexity classes and of the canonical sets representing them. To this end, a major tool is furnished by the resource-bounded reducibilities, which allow one to compare the "degree of difficulty" of the problems, and generate well-known mathematical structures (preorders) providing a clear and smooth way of defining and studying structural properties. Let us present some intuitive examples.

1. Maxima in classes defined by bounded resources (the so-called "complete" sets) are a central concept, which will receive major attention throughout this work.

2. Comparing reducibilities defined in different manners is a way of understanding better their structure, and it will be shown that some of the major problems of Complexity Theory can be considered as the still open relationship between the zero degrees of different reducibilities.

3. The open problems proposed by the transitivity of the nondeterministic reducibilities yield the possibility of defining hierarchies (analogous in some sense to the Kleene arithmetic hierarchy) whose different characterizations and properties furnish many intuitions about computation models.

4. A simple structural property, that of a set having a "slowly growing cardinality" (called "sparseness"), appears to be in a critical crossing of several pathways in the theory.

The reader will find many other examples throughout the book. We hope that those presented so far are sufficient to give the reader a global idea of the kind of results that (s)he can expect to find in this book. Undoubtedly, these results will improve his/her understanding of the concepts of resource-bounded computation, feasible and unfeasible algorithms, and realistically solvable and unsolvable problems.

1 Basic Notions About Models of Computation

1.1 Introduction

The aim of this chapter is to define many concepts that the reader is assumed to know, mainly in order to fix our notation. We present here some concepts of formal languages, set theoretic operations and relations, boolean formulas, and several models of computation, such as finite automata and the different versions of Turing machines: deterministic, nondeterministic, and oracle Turing machines.

The reader is advised that this first chapter does not aim to give a detailed exposition of the topics covered, but rather a setting of the notations and concepts which will be used freely throughout the book. To get a deeper background on the material covered in this chapter, we refer the reader to the bibliographical remarks at the end of the chapter.

We treat in a little more detail those topics that we consider important to this work but insufficiently explained in the currently available references. For example, our last section on oracle Turing machines is a bit more extensive than usual; in fact, oracle machines play crucial roles in many parts of the book, and we advise the reader unfamiliar with our model to read this part thoroughly for a better understanding of the remaining chapters.

1.2 Alphabets, Words, Sets, and Classes

We present in this section some basic concepts and definitions. For the interested reader who wishes to delve deeper into the subject, we refer to any of the standard books on foundations of theoretical computer science.

Definition 1.1
1. *An alphabet is any non-empty, finite set. We shall use upper case Greek letters to denote alphabets. The cardinality of alphabet Σ is denoted $|\Sigma|$.*
2. *A symbol, or also a letter, is an element of an alphabet.*
3. *Given an alphabet Σ, a chain, word, or even string over Σ is a finite sequence of symbols from Σ.*

4. *The length of a word w, denoted $|w|$, is the number of symbols w consists of.*

5. *The empty word λ is the unique word consisting of zero symbols.*

6. *For any word w over Σ and any symbol a in Σ, denote by $|w|_a$ the number of occurrences of the symbol a in w.*

7. *Given two words v and w over Σ, define the concatenation of v and w, denoted by vw (and sometimes by $v \cdot w$), as the word z which consists of the symbols of v in the same order, followed by the symbols of w in the same order.*

8. *A prefix of a word w over Σ is any word v such that for some word u, $w = vu$, i.e. any chain of symbols which form the beginning of w.*

9. *A suffix of a word w is any string v such that for some word u, $w = uv$, i.e. any chain of symbols which form the end of w.*

Notice that vertical bars are used with two different meanings: cardinality of alphabets (or other finite sets as indicated below), and length of strings. Let us define some easy operations on words.

Definition 1.2

1. *Given a word w on Σ and an integer n, define w^n inductively by: $w^0 = \lambda$, and $w^{n+1} = w \cdot w^n$ for all $n \geq 0$.*

2. *For any word w on Σ, define its reversal w^r inductively by: $\lambda^r = \lambda$, and $(u \cdot a)^r = a \cdot u^r$ for $a \in \Sigma$.*

Definition 1.3 *The set of all words over an alphabet Σ will be denoted Σ^*.*

The set Σ^* is a (non-commutative) monoid with the operation of concatenation, having λ as its only unit.

Given an alphabet Σ, we denote by Σ^n the set of all words over Σ with length n, and denote by $\Sigma^{\leq n}$ the set of all words over Σ with length less than or equal to n.

Most functions in this book are mappings from Σ^* into Γ^*, where Σ and Γ are alphabets. Composition of functions is denoted by the symbol $\circ$. Thus, $f_2 \circ f_1$ is the function obtained by applying first f_1 and then f_2. A special case of functions is given in the next definition.

Definition 1.4 *Given two alphabets Σ and Γ, a homomorphism is a mapping h from Σ^* to Γ^* such that*

1. $h(\lambda) = \lambda$ *and*
2. $h(u \cdot v) = h(u) \cdot h(v)$.

Every homomorphism is uniquely defined by the images of the letters of Σ. It can be extended to words in Σ^* by the equations: $h(\lambda) = \lambda$, and $h(w) = h(a)h(w')$ for $w = aw'$.

Using homomorphisms, we can consider different alphabets and make "translations" among them. Given two alphabets, Σ and Γ, words over Σ can be converted into words over Γ as follows: if Σ has fewer elements than Γ, then define any injective homomorphism such that the image of a letter of Σ is a letter of Γ; then words over Σ can be identified with words over a subset of Γ. If Σ has more elements than Γ, then choose a positive integer n such that $|\Sigma| \leq |\Gamma|^n$, and define an injective homomorphism mapping each element of Σ into a word of length n over Γ. (Observe that this requires that $|\Gamma| > 1$, since otherwise, for $|\Sigma| > 1$, there is no n fulfilling this condition.) The words over Σ are said to be *encoded* into words over Γ. To invert this encoding, retrieving the original word, decompose the word over Γ into subwords of length n and find the letters of Σ (if any) corresponding to these subwords. Notice that some words over Γ may encode no word over Σ.

Most of the objects we treat in this book can be described by words over some alphabet: boolean formulas, algorithms, Turing machines, instances to problems, etc. Using this scheme of encoding, we can identify any of these objects with a word over any alphabet of our choice. The only restriction is the indicated property that at least two letters are required.

This kind of encoding can also be applied to encoding several words over the same alphabet. Let us describe in more detail this operation, which will be extremely useful throughout this book. We want to encode several words into only one, in such a way that both computing the encoding and recovering the coded words can be easily done.

Given an alphabet Σ, choose a symbol "#" not in Σ, and associate the word $x\#y$ over the alphabet $\Sigma \cup \{\#\}$ to each pair of words x, y over Σ. Then fix a homomorphism h from $\Sigma \cup \{\#\}$ to Σ. The coding of the pair x, y is $\langle x, y \rangle = h(x\#y)$.

For example, take $\Sigma = \{0, 1\}$, and define h by: $h(0) = 00$, $h(1) = 11$, and $h(\#) = 01$. Then $\langle x, y \rangle$ is obtained by duplicating every bit of x, duplicating every bit of y, and appending the duplication of y to that of x, but inserting a 01 in between. Notice that the conversion defined by this homomorphism can be performed by a finite automaton with output, and that from the resulting word $\langle x, y \rangle$ the original words x and y can be recovered similarly by finite automata with output. Therefore this encoding is "as easy as possible", and every computation model able to simulate finite automata can compute and decode such encodings. This will be true for our standard model of computation, the Turing machine, which will be able to compute these functions using a minimum of resources: no more time than is needed to read the input, and no work space at all.

This kind of pairing function may not be bijective; there is no harm in that. If a non-bijective pairing function is being used, then just keep in mind that the decoding functions may be undefined on some inputs.

Pairing functions can be generalized to encode tuples as follows: $\langle x, y, z \rangle = \langle \langle x, y \rangle, z \rangle$; $\langle x, y, z, t \rangle = \langle \langle \langle x, y \rangle, z \rangle, t \rangle$; and so on. Even sequences of varying length can be encoded in this way, but then it is necessary to know the length of the sequence:

$$\langle x_1, x_2, \ldots, x_n \rangle = \langle \langle \langle \ldots \langle x_1, x_2 \rangle, \ldots \rangle, x_n \rangle, n \rangle$$

Given an alphabet Σ and an order of the symbols of Σ, we define an order on Σ^* in the following way: for each $n \geq 0$, all strings of length n are smaller than the strings of length $n + 1$; and within each length, the order follows a "lexicographic" criterion. This means that for any two different strings $w = a_1 a_2 \ldots a_m$ and $v = b_1 b_2 \ldots b_m$, of length m, w precedes v, denoted $w < v$, if there is a j with $1 \leq j \leq m$, such that for all $i < j$, $a_i = b_i$, but $a_j < b_j$.

This ordering is called the *canonical lexicographic order*.

Notice that the lexicographic order defines a bijection between the set $\mathbb{N} - \{0\}$ of the positive integers and the words in $\{0, 1\}^*$. This bijection can be described briefly as follows: the positive integer corresponding to the word w is obtained by prefixing w with a 1 (to avoid problems with leading zeros) and reading the result in binary. Conversely, given a positive integer, write it down in binary form and drop the leading 1 to obtain its associated word.

A similar bijection can be defined for any alphabet Σ, given an order on it. We can associate with each word w in Σ^* a positive integer which corresponds to the place occupied by w in the lexicographic order of Σ^*. Thus, Σ^* is always countably infinite.

We will consider sets of words over a given alphabet.

Definition 1.5 *Given a alphabet Σ, a language over Σ is a subset of Σ^*.*

We shall denote languages by roman upper case letters. The cardinality of a finite language A is denoted by $|A|$. Throughout this work, we shall refer frequently to languages as *sets* (of words). The empty language is denoted $\emptyset$.

For any language A on Σ, the definition of homomorphism can be extended in the following way:

$$h(A) = \{h(w) \mid w \in A\}$$

Let us define the notations we will use for some set theoretical operations on sets of words.

Definition 1.6 *Given languages A on Σ and B on Γ, define:*

1. *The union $A \cup B$ is the following language on $\Sigma \cup \Gamma$:*

$$A \cup B = \{w \mid w \in A \text{ or } w \in B\}$$

2. *The intersection $A \cap B$ is the following language on $\Sigma \cup \Gamma$:*

$$A \cap B = \{w \mid w \in A \text{ and } w \in B\}$$

3. *The concatenation $A \cdot B$ is the following language on $\Sigma \cup \Gamma$:*

$$A \cdot B = \{w \mid w = w_1 \cdot w_2 \text{ where } w_1 \in A \text{ and } w_2 \in B\}$$

4. *The difference of two languages A and B, both over Σ, is:*

$$A - B = \{x \mid x \in A \text{ but } x \notin B\}$$

5. *The symmetric difference of two languages A and B, both over Σ, is:*

$$A \triangle B = (A - B) \cup (B - A)$$

Notice that $x \in A \triangle B$ means that x is in A or in B but not in their intersection.

6. *The complement of a language A over Σ is the language $\overline{A} = \Sigma^* - A$.*

7. *Given Σ with at least two different symbols, say 0 and 1, the join or marked union of two languages A and B over Σ is:*

$$A \oplus B = \{w0 \mid w \in A\} \cup \{w1 \mid w \in B\}$$

Notice that if $\Sigma = \{0, 1\}$, reading in binary the words in $A \oplus B$ (preceded by 1, as indicated above) yields positive integers of the form $2 \cdot n$, where n is a word in A read in the same manner, or of the form $2 \cdot n + 1$, where n is a word in B read in the same manner.

It is an easy exercise to prove that if A and B are languages on Σ, then for any homomorphism h defined on Σ, $h(A \cup B) = h(A) \cup h(B)$ and $h(A \cdot B) = h(A) \cdot h(B)$.

Some properties of sets which will turn out to be useful in later chapters are related to the "density" concept. The *density* of a set is measured by the function indicating, for each integer n, the number of elements of length n or less contained in the set.

Definition 1.7   *A language is tally if and only if it is included in $\{a\}^*$ for some symbol a.*

Definition 1.8   *The census function of a set A over Σ is defined as follows: $C_A(n)$ is the number of words in A having length at most n.*

Definition 1.9   *We say that a set A is sparse if and only if its census function is bounded above by some polynomial, i.e. if there exists a polynomial $p(n)$ such that, for every n, $C_A(n) \leq p(n)$.*

This definition is equivalent to saying that a set A is sparse if there exists a polynomial $p(n)$ such that $|A \cap \Sigma^{\leq n}| \leq p(n)$.

The following properties are very easy to see:

Lemma 1.10   *Every tally set is sparse.*

Proof.   For any symbol a, every set A included in $\{a\}^*$ is sparse, since, for each n, there is at most one word with length equal to n in A. Therefore the total number of words of length less than or equal to n is at most $n+1$.   □

The converse is not true, as shown by the following easy counterexample:

Example 1.11   Take $\Sigma = \{0,1\}$. Let A be the set that, for each length n, consists of the smallest n words of this length according to the ordering of Σ^*:

$$A = \{0, 00, 01, 000, 001, 010, 0000, 0001, 0010, 0011, 00000 \dots \}$$

Then A contains n words of each length n, therefore it has a total of

$$\sum_{i=0}^{n} i = \frac{n \cdot (n+1)}{2} \leq n^2$$

words of length at most n. Therefore A is sparse.

Other densities are also interesting, and are used in Volume II (and less frequently than sparse sets).

Definition 1.12   *A set A has subexponential density if and only if for every nondecreasing polynomial $p(n)$, for every real constant $\varepsilon > 0$, and for all but finitely many n,*

$$C_A(p(n)) < \varepsilon \cdot 2^n$$

Thus, "subexponential density" means that the number of words of length up to $p(n)$ grows "strictly slower" than 2^n, in the sense that no constant factor is able to cover the gap. Observe that, assuming $p(n) \geq n$, this implies a bound on the number of words in A of *any* length, and not only of lengths that are of the form $p(n)$ for some n. Indeed, the number of words of any length n is smaller than the census at n, which is less than the census at $p(n)$, and therefore is bounded by $\varepsilon \cdot 2^n$.

There is also a dual concept:

Definition 1.13   *A set A has exponential density if and only if it is not the case that A has subexponential density; and a set A has symmetric density if and only if neither A nor $\overline{A}$ has subexponential density.*

It is immediately clear that A has exponential density if and only if there are real positive constants r_1 and r_2 such that for infinitely many n,

$$C_A(n^{r_2}) \geq r_1 \cdot 2^n$$

Given a subset A of Σ^*, its *characteristic function* or *indicator function* is the function χ_A from Σ^* to $\{0, 1\}$ defined by

$$\chi_A(x) = \begin{cases} 0 & \text{if } x \notin A \\ 1 & \text{if } x \in A \end{cases}$$

Notice that the characteristic function indicates the membership to a given set of any word over the alphabet.

For sets A whose words all have the same length n, we consider also the restriction of χ_A to Σ^n, and denote it also by χ_A if no confusion arises.

Languages over a given alphabet are interpreted as symbolic models of decisional problems. In an intuitive mood, a *decisional problem* can be described as a set of valid inputs (graphs, numbers, mathematical expressions, etc.) and a certain property that some of the inputs may satisfy (being connected, containing a Hamiltonian path, being a prime, satisfying certain equations, etc.). The problem consists in deciding whether a given valid input satisfies this property.

Suppose the problem is to be solved by any kind of algorithms. All models of the concept of algorithm work on symbolic inputs: most of the time, strings of characters; sometimes, integers; but never abstract concepts. Therefore, to state the problem in a way that algorithms may work on it, it is necessary to codify the valid inputs over some suitable alphabet. This is usually easy to do, because the "human" way of describing these instances suggests itself an encoding over some alphabet.

Now, given a problem, consider the set of valid inputs, encoded over some alphabet, and the subset of encodings of instances which satisfy the property defined by the decisional problem. Usually, there is no harm at all in considering jointly words encoding valid inputs which give a negative answer, and words that do not encode a valid input. Then the language of all the encodings of valid inputs that satisfy the property can be identified with the decisional problem, since deciding whether a given string belongs to this language is equivalent to deciding whether the input it encodes, if any, satisfies the property.

If the input is numbers, an obvious codification is obtained by expressing them in binary over the alphabet $\{0, 1\}$. Notice that the length of the input number n is $\lceil \log_2 n \rceil$. For more complicated structures, a larger alphabet usually helps to find a codification; then an appropriate homomorphism immediately furnishes an encoding over any other alphabet having at least two symbols, as indicated above.

The fact that homomorphisms can be computed by highly simple models of computation using a minimum of resources, such as finite automata, is meaningful: indeed, when we are studying problems in a class which is defined by bounding a given resource, the codification of the input must take no more resources than provided by the class under study.

We use terms like "class" or "family" to denote sets whose elements are sets of words. Unions of classes, intersections of classes, and other boolean operations are defined and denoted like their analogs for sets. The *boolean closure* of a class C is the smallest class containing C and closed under union, intersection, and complementation of sets in the class. One additional definition will be useful:

Definition 1.14 *Given a class C, we define its class of complements, denoted by co-C, by co-$C = \{L \mid \overline{L} \in C\}$.*

More generally, the notation "co-Prop" for a property "Prop" indicates that the property must hold for the complement of a set. For example, a *co-finite* set is the complement of a finite set, and a set is defined below to be "co-immune" if and only if its complement is "immune". The following easy properties hold:

Lemma 1.15 *Given classes of languages C_1 and C_2 over Σ^*, $C_1 \subseteq C_2$ if and only if co-$C_1 \subseteq$ co-C_2. In particular, $C_1 \subseteq$ co-C_1 if and only if $C_1 =$ co-C_1.*

Proof. Let $A \in$ co-C_1. Then $\overline{A} \in C_1$, which implies that $\overline{A} \in C_2$, and therefore $A \in$ co-C_2. The converse follows from the fact that co-co-$C = C$.
□

1.3 Inclusion Modulo Finite Variants

Several partial orders are used when dealing with complexity classes. Most of them arise from the concept of reducibility, which we shall define in Chapter 3. Another partial order is the inclusion modulo finite variants. Several results to be presented in this work are related to this partial order; for example, we shall see that many of the complexity classes we shall define are lattices under this order. We present in this section some definitions related to this concept.

Definition 1.16 *Given two sets A and B over some alphabet Σ, we say that A is a finite variation of B, or equivalently that A and B are equal almost everywhere (denoted $A = B$ a.e.), if and only if the symmetric difference $A \triangle B$ is a finite set.*

Notice that for sets A and B, the notation $A = B$ a.e. indicates that A and B coincide except for a finite set of elements.

It is straightforward to check that almost everywhere equality is an equivalence relation which induces, in the usual way, a partition of the subsets of Σ^* into equivalence classes. In our discussions, we will identify frequently the sets with their respective classes under this equivalence relation.

The inclusion modulo finite variants is a binary relation among sets which is appropriate to dealing with these equivalence classes.

Definition 1.17 *Set A is included in set B modulo finite variants if and only if all but finitely many of the elements of A are in B.*

It is also straightforward to check that inclusion modulo finite variants is a preorder, i.e. a reflexive and transitive binary relation, and that its associated equivalence relation is the almost everywhere equality. Most complexity classes we will study throughout this work are closed under almost everywhere equality. This property is defined as follows:

Definition 1.18 *A class $\mathcal{C}$ of sets is closed under finite variants (or also closed under finite variations) if and only if $A \in \mathcal{C}$ and $A = B$ a.e. implies $B \in \mathcal{C}$.*

Therefore, such a class is a union of (zero or more) equivalence classes, and the inclusion modulo finite variants is a well-defined partial order over it. For these classes, we often use the same symbol to denote both the class and its factorization modulo finite variations. Then the following proposition is straightforward.

Proposition 1.19 *Let $\mathcal{C}$ be a class of sets closed under union, intersection, and finite variants. Then $\mathcal{C}$ is a lattice under the partial order "inclusion modulo finite variants".*

There are bottom and top classes for this partial order, namely the class of the finite sets and the class of the co-finite sets. Given a class of sets $\mathcal{C}$, it is possible that for another fixed equivalence class $\mathcal{E}$ there is no equivalence class within $\mathcal{C}$, smaller than $\mathcal{E}$ in the partial order. This concept is formalized by the notion of "$\mathcal{C}$-immune" set, which is presented in the next definition together with other related properties.

Definition 1.20 *Let $\mathcal{C}$ be a class of sets.*

1. *A set L is $\mathcal{C}$-immune if and only if it is infinite and no infinite subset of L belongs to $\mathcal{C}$.*
2. *A set L is $\mathcal{C}$-co-immune if and only if its complement $\overline{L}$ is $\mathcal{C}$-immune.*
3. *A set is $\mathcal{C}$-simple if and only if it is in $\mathcal{C}$ and is $\mathcal{C}$-co-immune.*
4. *A set L is $\mathcal{C}$-bi-immune if and only if both L and $\overline{L}$ are $\mathcal{C}$-immune.*

Thus, a C-immune set A belongs to an equivalence class different from the class of finite sets; any class smaller than that of A corresponds to a subset B of A, and the immunity property implies that either B is finite, or B is not in C. On the other hand, observe that if L is bi-immune then both L and $\overline{L}$ are infinite. Let us see some examples.

Example 1.21 Fix the alphabet $\{0,1\}$, and let C be the class of all sets over the single letter alphabet $\{0\}$. An infinite set A is C-immune if and only if it contains only finitely many words consisting only of zeros, i.e. all but finitely many words of A contain at least one 1. A set B is C-simple if and only if it contains only words over $\{0\}$, and its complement is C-immune; i.e. if and only if B contains all but finitely many words over $\{0\}$. There are no C-bi-immune sets, because for every set A, either A or $\overline{A}$ (or both) contain infinitely many words consisting only of zeros.

Closure under finite variants is a concept that arises "almost everywhere" in this book. Immune, simple, and bi-immune sets are notions inherited from Recursive Function Theory, and will be of great importance in Volume II.

1.4 Boolean Formulas

In this section we review some basic concepts of propositional logic, and introduce a terminology which will be used frequently in this book.

Definition 1.22 *A boolean variable is any symbol to which we can associate either of the truth values 0 and 1 (sometimes denoted also by "false" and "true").*

Definition 1.23 *Let X be a countable set of boolean variables. The class of the boolean formulas over X is the smallest class which is defined by:*

1. *The boolean constants 0 and 1 are boolean formulas.*
2. *For every boolean variable x in X, x is a boolean formula.*
3. *If F_1 and F_2 are boolean formulas then $(F_1 \wedge F_2)$, $(F_2 \vee F_2)$, and $\neg(F_1)$ are boolean formulas.*

Furthermore, the class of the quantified boolean formulas over X is obtained by the rules:

4. *Every boolean formula is a quantified boolean formula.*
5. *If x is in X and F is a quantified boolean formula, then $\exists x F$ and $\forall x F$ are quantified boolean formulas.*

The symbols $\wedge$, $\vee$, and $\neg$ are called *connectives*. We use the same symbols for the concepts of "and", "or", and "not". The symbols $\exists$ and $\forall$ are, respectively, the *existential quantifier* and the *universal quantifier*.

Given a quantified boolean formula, a variable which occurs in the formula but is not affected by a quantifier is called a *free* variable.

Notation. Let F be any quantified boolean formula on X, x a variable in X, and $a \in X \cup \{0,1\}$. Then $F|_{x:=a}$ is the formula obtained by substituting a for every free occurrence of x in F. Notice that the given syntax implies that either none or all the occurrences of a variable in a formula are free. If a itself is a variable, then this process is called *renaming of variables*.

We assume we have a well-defined infinite set of different variable symbols X, which is fixed from now on. Unnecessary parentheses will be omitted when writing down boolean formulas.

Definition 1.24   *A boolean assignment is a mapping from X to $\{0,1\}$. Given a quantified boolean formula F and an assignment V, the value of F under V is the boolean value defined as follows:*

1. $V(F) = 0$ *if* $F = 0$
2. $V(F) = 1$ *if* $F = 1$
3. $V(F) = V(x)$ *if* $F = x$ *for* $x \in X$
4. $V(F) = V(F_1) \vee V(F_2)$ *if* $F = (F_1 \vee F_2)$
5. $V(F) = V(F_1) \wedge V(F_2)$ *if* $F = (F_1 \wedge F_2)$
6. $V(F) = \neg V(F_1)$ *if* $F = \neg(F_1)$
7. $V(F) = V(F_1|_{x:=0}) \vee V(F_1|_{x:=1})$ *if* $F = \exists x F_1$
8. $V(F) = V(F_1|_{x:=0}) \wedge V(F_1|_{x:=1})$ *if* $F = \forall x F_1$

Observe that the value of F only depends on the value of x if x is free in F, and that this does not hold for the last two cases in this definition.

Definition 1.25   *Given a boolean formula F, we say that an assignment V satisfies F if $V(F) = 1$. A formula F is satisfiable if and only if there is an assignment V which satisfies F; otherwise F is unsatisfiable.*

Example 1.26   Let $X = \{x_1, \ldots, x_n\}$ be the set of boolean variables, and fix $r < n$. Then the formula

$$x_1 \wedge x_2 \wedge \cdots \wedge x_{n-r}$$

is satisfied by all assignments giving the value 1 to the first $n - r$ variables, and giving arbitrary values to the remaining variables. The total number of satisfying assignments is 2^r.

Example 1.27   Let $X = \{x_1, \ldots, x_n\}$ be the set of boolean variables. Using the idea of the previous example, we construct a formula having exactly i satisfying assignments, for any given $i < 2^n$. By computing the binary representation of i, find k integers ($n \geq r(1) > r(2) > \cdots > r(k) \geq 1$) such that

$$i = 2^{r(1)} + 2^{r(2)} + \cdots + 2^{r(k)}$$

and construct the formula:

$$(x_1 \wedge x_2 \wedge \cdots \wedge x_{n-r(1)}) \vee$$

$$(x_1 \wedge \cdots \wedge \neg x_{n-r(1)} \wedge \cdots \wedge x_{n-r(2)}) \vee$$

$$(x_1 \wedge \cdots \wedge \neg x_{n-r(1)} \wedge \cdots \wedge \neg x_{n-r(2)} \wedge \cdots \wedge x_{n-r(3)}) \vee$$

$$\cdots$$

$$\cdots$$

$$\vee (x_1 \wedge \cdots \wedge \neg x_{n-r(1)} \wedge \cdots \wedge \neg x_{n-r(k-1)} \wedge \cdots \wedge x_{n-r(k)})$$

Each of the conjunctions corresponds to one of the terms in the decomposition of i: the jth conjunction contributes with $2^{n-r(j)}$ satisfying assignments. Furthermore, the negations ensure that no assignment satisfying one of the conjunctions can satisfy another, so that we are sure that no assignment is counted twice. Therefore the total number of satisfying assignments is

$$2^{n-r(1)} + 2^{n-r(2)} + \cdots + 2^{n-r(k)} = i$$

This example will be used later in Chapter 6.

Definition 1.28   *Given two quantified boolean formulas F_1 and F_2, we say that they are equivalent, and denote this by $F_1 \equiv F_2$, if and only if, for every assignment V, $V(F_1) = V(F_2)$.*

The following useful results are quite easy to prove and are left to the reader. From now on, they will be assumed throughout the text.

Lemma 1.29   *Let F_1, F_2, and F_3 be boolean formulas. The following statements are true:*

(a) $\neg\neg F \equiv F$
(b) $\neg(F_1 \wedge F_2) \equiv (\neg F_1 \vee \neg F_2)$
(c) $\neg(F_1 \vee F_2) \equiv (\neg F_1 \wedge \neg F_2)$
(d) $(F_1 \vee F_2) \equiv (F_2 \vee F_1)$
(e) $(F_1 \wedge F_2) \equiv (F_2 \wedge F_1)$
(f)   *If $F_1 \equiv F_2$ then $\neg F_1 \equiv \neg F_2$*
(g)   *If $F_1 \equiv F_2$ then $F_1 \wedge F_3 \equiv F_2 \wedge F_3$*
(h)   *If $F_1 \equiv F_2$ then $F_1 \vee F_3 \equiv F_2 \vee F_3$*

Lemma 1.30 *Let F_1 and F_2 be quantified boolean formulas, and let $F_1 \equiv F_2$. Then the following statements are true:*

1. $\exists x F_1 \equiv \exists x F_2$
2. $\forall x F_1 \equiv \forall x F_2$

Proposition 1.31 *Let F be a quantified boolean formula, and y a variable not in F. Then for any boolean variable x:*

$$\exists x F \equiv \exists y F|_{x:=y} \quad and \quad \forall x F \equiv \forall y F|_{x:=y}$$

Proof. We prove the existential case. The other case is similar. Let V be any assignment. Then:

$$
\begin{aligned}
V(\exists x F) &= V(F|_{x:=0}) \vee V(F|_{x:=1}) = \\
&= V(F|_{x:=y}|_{y:=0}) \vee V(F|_{x:=y}|_{y:=1}) = V(\exists y(F|_{x:=y}))
\end{aligned}
$$
$\square$

The following definitions allow us to present boolean formulas in a particular and useful form.

Definition 1.32 *A literal is a boolean formula which is either a boolean variable, or the negation of a boolean variable. A clause is a disjunction of literals.*

Thus, a clause is a formula of the form

$$C = \ell_1 \vee \ell_2 \vee \cdots \vee \ell_n$$

where each ℓ_i is either a boolean variable x_k, or its negation $\neg x_k$.

Now we can define the conjunctive normal form (also called clausal form) of boolean formulas:

Definition 1.33 *A boolean formula is in conjunctive normal form (CNF for short) if and only if it is the conjunction of a set of clauses.*

Thus, F is in CNF if it is of the form

$$F = C_1 \wedge C_2 \wedge \cdots \wedge C_n$$

where each C_k is a clause as before.

Boolean formulas with or without quantifiers can be written down over suitable alphabets, as we have been doing up to now. For instance, we can encode a boolean formula as a word over the alphabet

$$\{\wedge, \vee, \neg, \forall, \exists, (,), 0, 1, false, true\}$$

where each variable is represented by the binary expression of its subindex, and *false* and *true* denote the boolean constants. Whenever it is convenient to use a different alphabet, an encoding scheme can be used as described in the previous sections.

Definition 1.34   *Fix an alphabet Σ, and assume that formulas are encoded into words over Σ. Some important languages over this alphabet, related to boolean formulas, are:*

1. SAT, *the set of all encodings of satisfiable quantifier-free boolean formulas;*
2. CNF-SAT, *the set of all encodings of satisfiable quantifier-free boolean formulas in conjunctive normal form; and*
3. QBF, *the set of all encodings of quantified boolean formulas, without free variables, which evaluate to "true".*

The size of a formula is the length of the word encoding it over some fixed alphabet. Observe that the number of occurrences of variables is linear in the number of connectives. If a formula contains n variables, then each variable can be written down with $\lceil \log n \rceil$ symbols. Thus, the size of a formula is a constant times $m \cdot \log m$, where m is the number of connectives.

Hence, to abstract from the alphabet used in our arguments about the size of a formula, we shall argue about the total number of connectives in it.

Theorem 1.35   *For each boolean formula F having m connectives, with boolean variables $x_1, \ldots, x_t$, there exists an equivalent quantified boolean formula*

$$\exists y_1 \exists y_2 \ldots \exists y_k F'$$

where in F' there occur just the variables $x_1, \ldots, x_t, y_1, \ldots, y_k$, such that F' is a boolean formula in CNF having $c \cdot m$ connectives for some constant c independent of F.

Proof.   First note that each boolean formula is equivalent to a formula in which the negation $\neg$ is only applied to variables, by repeated application of Lemma 1.29. Thus, suppose that F contains only literals, conjunctions, and disjunctions. We prove the lemma by structural induction on the form of the formula. The construction may give formulas with constants, which are not literals. We then describe how to remove the constants.

F *is a literal.* This case is immediate: no transformation is required at all, and no new variables are added.

F *is a constant.* This case is also immediate: leave F unchanged.

$F = F_1 \wedge F_2$. By the induction hypothesis, and after a renaming of variables, there are formulas $\exists y_1 \ldots F_1'$ and $\exists z_1 \ldots F_2'$ which fulfill the conditions and are equivalent respectively to F_1 and F_2. Form F' as the conjunction of all the clauses in F_1' and F_2'. Then $\exists y_1 \ldots \exists z_1 \ldots F'$ is equivalent to F. Renaming again the variables we obtain a formula satisfying the conditions.

$F = F_1 \vee F_2$. This is the only case that requires addition of variables. Again by induction hypothesis and after renaming of variables, there are formulas $\exists y_1 \ldots F_1'$ and $\exists z_1 \ldots F_2'$ which fulfill the conditions and are equivalent respectively to F_1 and F_2. Observe that F is true if and only if either F_1 is true or F_2 is true. Let y be a new variable; intuitively, y will be used to select whether F is true because of F_1 or because of F_2. In a certain sense, if y is assigned 0, then F is satisfied if and only if F_1 is satisfied; and if y is assigned 1, then F is satisfied if and only if F_2 is satisfied. Therefore y will be quantified existentially.

In order to achieve this, construct new formulas as follows. From F_1', construct F_1'' by converting each clause C in F_1' into a new clause $(y \vee C)$ in F_1''. From F_2', construct F_2'' by converting each clause C' in F_2' into a new clause $(\neg y \vee C')$ in F_2''. Now form the formula F' as the conjunction of all the clauses in F_1'' and F_2''. Then $\exists y \exists y_1 \ldots \exists z_1 \ldots F'$ fulfills the conditions and is equivalent to F, as can easily be seen.

Finally, remove the boolean constants as follows: if any clause contains the constant 0, remove the constant from the clause; and if any clause contains the constant 1, remove the whole clause, since it is trivially satisfied. A renaming of variables completes the construction.

It is easy to see that the size of the formula obtained is $c \cdot m$ for some constant c. This completes the proof. $\qquad \Box$

Corollary 1.36 *For each boolean formula F having m connectives, there exists another boolean formula F' in CNF having $c \cdot m$ connectives for some constant c independent of F, and depending on more variables, such that F is satisfiable if and only if F' is satisfiable.*

Proof. Form the formula $\exists y_1 \ldots F'$ as in the previous theorem, and just drop all the quantifiers. The formula F' is satisfiable if and only if F is. $\quad \Box$

1.5 Models of Computation: Finite Automata

For the sake of completeness, let us define some basic models of computation which the reader should already be familiar with. Otherwise we refer to any of the basic texts on formal language theory indicated in the bibliographical remarks at the end of this chapter.

Definition 1.37 *A non-deterministic finite automaton is a five-tuple*

$$\mathcal{A} = \langle Q, \Sigma, \delta, Q_-, Q_+ \rangle$$

where:

1. *Q is a finite non-empty set, whose elements are called states;*

2. Σ *is a finite alphabet;*
3. $\delta : Q \times \Sigma \to \mathcal{P}(Q)$ *is the transition function. Here $\mathcal{P}(Q)$ denotes the set of all subsets of Q. Hence, for each pair (state, symbol), δ specifies a set of legal new states;*
4. $Q_- \subseteq Q$ *is the set of initial states;*
5. $Q_+ \subseteq Q$ *is the set of final states.*

When $|\delta(q, a)| \leq 1$ for every state q and symbol a, the transition function can be identified with a partial function $\delta : Q \times \Sigma \to Q$. If, moreover, $|Q_-| = 1$, then we say that the automaton is *deterministic*.

Finite automata may *accept* or *reject* words over Σ. This is defined as follows:

Definition 1.38 *For any given finite automaton $\mathcal{A}$, a computation of $\mathcal{A}$ on a word $w = a_1 a_2 \ldots a_n$ is a sequence $(q_0, a_1, q_1), (q_1, a_2, q_2), \ldots, (q_{n-1}, a_n, q_n)$, where $q_0 \in Q_-$, and for each triple (q_i, a_{i+1}, q_{i+1}), it holds that $q_{i+1} \in \delta(q_i, a_{i+1})$.*

Definition 1.39 *Given a word $w = a_1 a_2 \ldots a_n$ and an automaton $\mathcal{A}$, w is accepted by $\mathcal{A}$ if and only if there is a computation on w which ends in a final state $q_n \in Q_+$. Otherwise w is rejected. The language accepted by $\mathcal{A}$ is the set of all words accepted by $\mathcal{A}$.*

1.6 Models of Computation: Turing Machines

In this section we shall describe the *deterministic Turing machine* model. Other variants of Turing machines are defined in the remaining sections of this chapter.

Our concept of Turing machine corresponds to that of "multitape off-line" Turing machine (or sometimes, as we shall see, that of "multitape on-line" Turing machine). This is a device which consists of:

1. A finite set of states as in the case of finite automata, sometimes also called the *finite control* of the machine;
2. one or more (but a fixed number of) semi-infinite *work tapes* divided into *cells*, each equipped with a *tape head* which can move right or left, scanning the cells of the tape, one at a time; and
3. an extra tape with its corresponding head, called the *input tape*.

The term "semi-infinite tape" means that it has no rightmost cell, but it does have a leftmost cell. When the head is on the leftmost cell, it is not allowed to move left.

At each moment, the device is in one of the states. Then the device can read the contents of the scanned cells of all tapes, change the contents of the scanned cells of the work tapes by writing new symbols on them, move each head right or left, and change its internal state. All these operations form a *step*, and are uniquely defined by a *transition function*, as a function of the current internal state and of the symbols read from the tapes. Observe that it is not allowed to change the contents of the input tape.

A machine M starts operating on an input word w with the input tape containing w, and a fixed symbol called *blank* in the remaining cells. Every cell of the work tapes also contains a blank. The internal state is the initial state q_0. Then M proceeds by applying the transition function as long as possible. Whenever the transition function is undefined, the machine stops. If after a sequence of steps, the machine stops in a final accepting state, then we say that M *accepts* w, otherwise M *rejects* w. Observe that there are two ways of rejecting: either by stopping in a non-accepting state, or by never stopping.

Let us now present a more formal definition:

Definition 1.40   *A Turing machine with k tapes is a five-tuple*

$$M = \langle Q, \Sigma, \delta, q_0, F \rangle$$

where:

1. *Q is the finite set of internal states;*
2. *Σ is the tape alphabet;*
3. *$q_0 \in Q$ is the initial state;*
4. *F is the set of accepting final states, and*
5. *$\delta : Q \times \Sigma^k \to Q \times \Sigma^{k-1} \times \{R, N, L\}^k$ is a partial function called the transition function of M.*

Situations in which the transition function is undefined indicate that the computation must stop. Otherwise the result of the transition function is interpreted as follows: the first component is the tuple of symbols to be written on the scanned cells of the $k - 1$ work tapes; the second component is the new state, and the third component specifies the moves of the tape heads: R means moving one cell to the right, N means do not move, L means moving one cell to the left if there are cells to the left, otherwise do not move.

We now introduce the concepts of "configuration" and "computation" of Turing machines.

Definition 1.41   *Given a machine M, a configuration of M, also called an instantaneous description, or also a snapshot, is just a description of the whole status of the computation: it includes the contents of each of the tapes, the position of each of the tape heads, and the state of the Turing machine program in the finite control. If M has k tapes, a configuration of M is a $k + 1$ tuple*

$$(q, x_1, x_2, \ldots, x_{k-1}, x_k)$$

where q is the current state of M, and each $x_j \in \Sigma^* \# \Sigma^*$ represents the current contents of the jth tape. The symbol "$\#$" is supposed not to be in Σ, and marks the position of the tape head (by convention, the head scans the symbol immediately at the right of the "$\#$"). All symbols in the infinite tape not appearing in x_j are assumed to be the particular symbol "blank".

At the beginning of a computation, all tapes are blank except the input tape.

Definition 1.42 *The initial configuration of a machine M on an input w is $(q_0, \#w, \#, \ldots, \#)$.*

The fact of accepting an input will be indicated by reaching an "accepting configuration".

Definition 1.43 *An accepting configuration is a configuration whose state is an accepting state.*

By our convention regarding the input tape, its contents cannot change during the computation of a machine on a given input. Therefore, when the input is fixed, there is no need to include the contents of the input tape in the configuration: just the position of the input head is required.

Definition 1.44 *Given a machine M and a fixed input w, a configuration of M on input w is a configuration in which the contents of the input tape is omitted, and only the position of the input tape head is recorded (in binary).*

Thus, a configuration on a fixed input has the form

$$(q, i, x_2, \ldots, x_{k-1}, x_k)$$

where i is the position of the input tape head and the other parameters are as described above.

A "computation" can now be defined as a sequence of configurations:

Definition 1.45 *Given a machine M and an input string w, a partial computation of M on w is a (finite or infinite) sequence of configurations of M, in which each step from a configuration to the next obeys the transition function. A computation is a partial computation which starts with the initial configuration of M on w, and ends in a configuration in which no more steps can be performed.*

Of course, an *accepting computation* is a computation ending in an accepting configuration, and in this case the input word is *accepted*. The *language accepted* by the machine M, denoted $L(M)$, is the set of words accepted by M.

Often in this book we will use the finite control of the machines to maintain a finite amount of information. More precisely, we use this method to modify the behavior of a machine on a given finite set of "exceptional" inputs. The method amounts to incorporating into the program a built-in "table" containing this finite set of inputs, and for each one an indication whether to accept or to reject it.

Formally, this is done as follows. A deterministic finite automaton is constructed which accepts exactly the finite set of exceptions; it has the form of a tree, each exception corresponding to a branch of the tree. The final states are marked, indicating whether the machine must accept or reject each exception. A new machine is constructed which starts simulating the automaton. If a final state of the finite automaton is reached, it indicates whether the new machine should accept or reject; if during the computation the automaton finds out that the input is not an exception, the input head comes back to the left end and the computation begins again as would be done by the old machine. This kind of transformation of a machine will be referred to by expressions such as *handling a finite number of exceptions by table look-up* or *finite "patching" of the machine*.

Sometimes we will wish to use Turing machines to compute functions. The appropriate model is obtained by adding to the Turing machine an output tape, usually imposing the condition that its tape head may only write on it and move one cell to the right after writing. Such a machine is usually called a *transducer*. When it must be clear that a machine is *not* a transducer, sometimes the name *acceptor* is used.

If the machine is a transducer, the function it computes is defined on its accepted language, and for each accepted input the value of the function is the word which appears in the output tape when the machine stops in an accepting state.

The concepts of configuration and computation for transducers are defined exactly as for acceptors. The description of the output tape is included in the configuration. '

Let us discuss briefly how to encode machines, configurations, and computations over fixed alphabets.

To describe a machine, there must be a way of writing it down. For example, given a Turing machine with k tapes, $M = \langle Q, \Sigma, \delta, q_0, F \rangle$, we can codify it by encoding each legal transition over the alphabet $\{0, 1\}$. One possible way of doing this is to write each of the entries of the transition function over the alphabet Γ which contains the symbols "0", "1", "δ", "(", ")", "=", "R", "N", "L", and "," as follows: previously, enumerate the set Q of states and the symbols of Σ (this must be done so that the encoding is valid for arbitrarily large machines with arbitrarily large alphabets). Then each entry of the transition function takes the format

$$\delta(q, x_1, \ldots, x_k) = (y_1, \ldots, y_{k-1}, q', Z)$$

where q, q', x_i, and y_i are words of $\{0,1\}^*$ interpreted as numbers of states or of symbols, and $Z \in \{R, N, L\}^*$. Now the machine M is a word over Γ formed by the concatenation of all entries of its transition function, and it can be translated into any other alphabet having at least two symbols via a homomorphism as described in Section 1.2.

As a concrete example, Γ may be translated into $\{0,1\}^*$. Then each machine is codified as a string over $\{0,1\}$, and the set of all encodings of machines will form a language over $\{0,1\}$. Thus, according to the remarks made in the first section of this chapter, to each machine correspond an integer, and to some integers corresponds a machine. Furthermore, the machine can be recovered easily from the integer. This process is known as a *gödelization* of Turing machines, and the number corresponding to each machine is its *Gödel number*. By convention, numbers that do not correspond to a string correctly encoding a machine are considered Gödel numbers of some arbitrarily fixed trivial machine. In this way we can list in order all the Turing machines, and refer to the nth machine, meaning the Turing machine with Gödel number n.

Configurations and computations can be described as well by words over some alphabet, and we will identify configurations and computations with the words describing them. Our conventions about translation between alphabets also apply here.

All these remarks and definitions for Turing machines apply too to the case of transducers: we can talk about the Gödel number of a transducer, a configuration of a transducer, or a computation of a transducer.

The reader should recall that it is possible to design a Turing machine, called the *universal machine*, which receives as input the encoding of a machine M together with an input w to M, and which is able to simulate the computation of M on w. This fact is similar to the existence of "interpreters" for computer programming languages, i.e. programs that read and execute other programs.

Let us also say here a couple of words about variants of Turing machines. Occasionally, we will use *on-line* Turing machines. The only difference from the model described above (which is referred to as an *off-line* Turing machine) is that in on-line Turing machines the transition function is not allowed to specify the move L for the input tape head. Therefore, in these machines the input can be read only forwards, and once the input tape head has left a cell it cannot backtrack to read this cell again.

This model raises a problem for the finite patching described above, since it cannot re-initialize itself after checking for an exception. A possible solution is as follows: the modified machine has as states pairs formed by a state of the old machine and a state of the automaton. When the input tape head scans a new symbol of the input, the state of the automaton is modified according to its transition function; the state of the old machine simulates the computation of the old machine. In this way, finitely many exceptions can be "patched" into the on-line machines.

Another model which appears occasionally is the single tape Turing machine. In this model there are no work tapes, and the input tape head is allowed to move in either direction and to change the symbols on its single tape.

There exist many other modifications of the Turing machines which are equivalent to the basic model described above, in the sense that they accept the same class of languages (the recursively enumerable sets). None is used in this book. We list some of them here: Turing machines with two-sided infinite tapes; Turing machines with several heads on each tape; and Turing machines with multidimensional "tapes". Proofs of the equivalence can be found in the textbooks listed in the references.

There are many other models of computation which are equivalent to the Turing machines; some of them are highly different from them. Their abundance suggests that Turing machines correspond to a natural, intuitive concept of "algorithm".

Other natural computational models can be devised which define different classes (such as the oracle machines defined later, or the nonuniform models presented in Chapter 5). These concepts usually have some clear difference from our concept of algorithm: for instance, some steps must be performed as indicated by an external agent, or sometimes the algorithm is expected to be applied only to a finite set of potential input data. However, the naive properties of mechanicity and applicability to data of unbounded size seem to correspond to the class of algorithms defined by the Turing machines.

Thus, the following "axiom" will be accepted henceforth: *every algorithm can be described by a Turing machine*. This principle is widely known as *Church's Thesis*.

The Turing machine model of computation is not friendly. Presenting algorithms as Turing machines is usually a tedious and unrewarding task. Turing machine programs are difficult to write and very hard to understand, and arguing about the correctness or about the amount of resources needed to perform a particular Turing machine algorithm is similar to doing it for executable binary machine code. Therefore, throughout this work, whenever a concrete algorithm is to be presented, we shall appeal to Church's Thesis, and describe it in high-level, human oriented pseudo-code close to a PASCAL style. Our algorithmic language is full of ambiguities which, however, can always be solved easily using contextual information. Let us present an example.

Example 1.46 Given a boolean formula F on n variables, we wish to decide whether F is satisfiable; i.e., we want to decide the set SAT. The algorithm is presented in Figure 1.1.

```
input F
check that F encodes a syntactically correct boolean formula
let n be the number of different variables in F
for each sequence w in {0, 1}ⁿ do
    for i := 1 to n do
        substitute the iᵗʰ variable in F by 0 or 1,
            according to the iᵗʰ bit of w
        simplify the resulting formula
        if the result is "true"
            then halt in an accepting state
end for's
halt in a rejecting state
end
```

Figure 1.1   An algorithm for deciding SAT

A question that arises immediately is the following: since Turing machines are not to be used for presenting algorithms, why have we chosen them as our model of computation? The answer is the following: first, this model of computation offers a clear, simple, and unambiguous definition of a computation step, and hence of time required by a computation. Second, this model also offers a clear, simple, and unambiguous definition of a storage unit, the tape cell (or, more formally, each symbol from the tape alphabet), and hence of the storage employed in a computation. This is not the same for every computation model.

Furthermore, these two concepts are in some sense realistic. It is a fact that for every two computation models taken from a wide family of reasonable computing devices, they simulate each other with an at most quadratic time overhead and within a constant factor of space. Multitape Turing machines are a prominent member of this class. (An interesting family of models which seem to be more powerful within the same resources are the parallel models to be studied in Volume II.) Moreover, this property is required in the definition of all the complexity classes central to our study, since it makes them independent of the computation model used. Thus, it can be admitted (and we explicitly admit it) as an extension of Church's Thesis, that this fact holds for all "reasonable" models of sequential computation. This extension has been called the *Invariance Thesis*. It is not our aim here to study models of computation, nor to show their equivalence with multitape Turing machines. Nevertheless, we should point out that this equivalence does not hold for some interesting models of computation.

1.7 Models of Computation: Nondeterministic Turing Machines

In the model presented in the previous section, every move is completely determined by the current situation. The state of the machine, and the symbols currently scanned by the tape heads, completely determine the next state and the moves of the tape heads.

In this section we relax this condition to obtain nondeterministic devices, whose "next move" can be chosen from several possibilities. Formally, a nondeterministic Turing machine is defined as a five-tuple

$$M = \langle Q, \Sigma, \delta, q_0, F \rangle$$

where each element is as in the deterministic case, with the exception that the transition function is defined by

$$\delta : Q \times \Sigma^k \to \mathcal{P}(\Sigma^{k-1} \times Q \times \{R, N, L\}^k)$$

where, for any set A, $\mathcal{P}(A)$ denotes the power set of A.

Of course, all the definitions and remarks made for the deterministic case about encodings of machines, configurations, and computations, apply in the same manner to the nondeterministic model. However, on a given input there is now not only one computation, but a set of possible computations. Acceptance for nondeterministic machines is therefore defined as follows.

Definition 1.47 *An input w is accepted by a nondeterministic machine M if and only if there exists a computation of M on w ending in an accepting configuration. We denote by $L(M)$ the language accepted by M, i.e. the set of all words accepted by M.*

If use of resources is not considered, then nondeterministic machines have the same power as deterministic ones, and hence fulfill Church's Thesis. However, it is not true that the power remains the same under a limitation of the resources. Most of this book is devoted to understanding this fact. Other modifications of the nondeterministic Turing machines which are equivalent to our model arise from the same considerations given for deterministic machines: number of tapes, number of heads, multidimensional "tapes",.... We do not discuss them here.

Observe that allowing a constant number of "next move" possibilities greater than two is in some sense useless. An election among k possibilities can be substituted by $\lceil \log k \rceil$ elections among two possibilities. If k is a constant, then the computation is only a constant factor longer. Thus, we lose no generality if we require in our model that for every pair (q, a), $|\delta(q, a)| \leq 2$. This fact is usually expressed by saying that the machines have "nondeterministic fan-out 2".

Given this condition on the nondeterministic fan-out, the set of possible computations of a nondeterministic machine M on an input string w can be described by a binary *tree of computations* in the following manner: the nodes of the tree are configurations of the machine M on input w; the root is the initial configuration, and for any node c, its sons are those configurations which can be reached from c in one move according to the transition function of the machine. Leaves of the tree are final configurations, some of which may accept. An accepting computation is a path starting at the root and finishing at an accepting leaf. By the definition of acceptance given above for nondeterministic machines, the input is accepted if and only if there is at least one accepting leaf in the tree.

Thus, at each internal node, the selection of the continuation of the path is done in a nondeterministic manner. Each internal node represents a binary "guess" of how to follow correctly one of the accepting paths. This is also equivalent to labeling each internal node with a "disjunction": the node leads to acceptance if and only if either the successor at the left, or the successor at the right, leads to acceptance.

A very general nondeterministic scheme which appears often in this book is presented in Figure 1.2. We use the following notation for it.

Notation. The nondeterministic process of Figure 1.2 will be abridged to *guess y with $|y| \leq i$*.

$y := \lambda$
for i steps do
      choose in a nondeterministic manner
           $y := y0$ or
           $y := y1$ or
           keep y without change

Figure 1.2   Meaning of *guess y with $|y| \leq i$*

The computation tree of a nondeterministic machine on input w could be an infinite tree. However, any accepting path must be a finite path, since it ends in a final, accepting configuration. We will use this fact later as follows: if we know that a machine must accept within a given, previously known number of steps, say t, then every computation longer than this number can be aborted, since this computation cannot be the computation accepting in t steps. In other words, we prune the computation tree at depth t. In this way, a potentially infinite computation tree can be transformed into a finite tree without modifying the accepted set.

Example 1.48 In Figure 1.3 we present a nondeterministic algorithm for the satisfiability problem of Example 1.46.

---

input F
check that F encodes a correct boolean formula
for each variable x occurring in F do
    choose in a nondeterministic manner
        $F := F|_{x:=0}$ or
        $F := F|_{x:=1}$
simplify the resulting formula
if the result is "true" then accept and halt
end

Figure 1.3 A nondeterministic algorithm for SAT

The differences between the deterministic algorithm of Example 1.46 and the nondeterministic algorithm of Example 1.48 are easy to see. The deterministic algorithm may have to explore all the possible assignments of 0 and 1 to the variables. However, in the nondeterministic model, the guessing is made only once, and we only have to find what is the resulting value of F. In this sense the application of nondeterminism is equivalent to first guessing the solution, and then testing that it indeed works.

Other models of computation in which the next step is not uniquely defined will be studied later. They differ from nondeterministic machines in the definition of accepted word. These include probabilistic machines (studied in Chapter 6) and alternating machines (studied in Volume II).

1.8 Models of Computation: Oracle Turing Machines

In order to compare later the degree of difficulty of solving a problem, we will allow our algorithms to operate with the help of some external information. One of the ways of doing this is presented in this section; others will appear later in this book.

Definition 1.49 *An oracle machine is a multitape Turing machine M with a distinguished work tape, called oracle tape or query tape, and three distinguished states QUERY, YES and NO.*

The computation of an oracle machine requires that a set, called *oracle set*, be fixed previously to the computation. At some step of a computation on input string x, M may transfer into the state QUERY. From this state, M transfers into the state YES if the string currently appearing on the query tape belongs to the oracle set; otherwise, M transfers into the state NO. In either case the query tape is instantly erased, and no other change occurs in the configuration of the machine. States YES and NO are sometimes called *answer states*. We call such a query step just a *query*, and we refer to the word written in the oracle tape in a query step as the *queried word*. Sometimes oracle machines are said to compute *relative to* their oracle sets.

Of course, non-query steps may appear in the computation, and proceed exactly as in the case of non-oracle Turing machines. Therefore, oracle machines may be deterministic or nondeterministic. An example of an algorithm representing an oracle machine is presented below (Figure 1.4).

The definition of configuration extends to the oracle machine in a straightforward way, as does the definition of computation of an oracle machine M on input w with oracle A. Accepting computations and accepted words are also defined in the same way. Given an oracle machine M and an oracle A, we denote by $L(M, A)$ the *language accepted* by an oracle machine M relative to oracle A, i.e. the set of all words accepted by M with oracle A.

Encodings of machines, configurations, and computations are defined as for the non-oracle case, and these encodings can be interpreted as integers (their Gödel numbers) by translation into the alphabet $\{0, 1\}$. This provides an enumeration of the oracle Turing machines.

Non-oracle Turing machines can be considered as oracle machines in a straightforward way. We will identify oracle machines using the empty oracle set with non-oracle machines whenever it is adequate.

We can also associate a tree of computations to the computations of oracle machines, but defined in a very different manner than for the nondeterministic machine. Given an oracle machine M, whose oracle set is not yet fixed, and an input string x, the *tree of computations of M on x* is a binary tree in which the root of the tree corresponds to the initial configuration of M on x; it has just one son, corresponding to the configuration in which the first query is made, labeled with the queried word. Similarly, every internal node other than the root represents a query of M to the oracle, and is labeled with the queried string.

Every left branch from an internal node is labeled "yes" and every right branch is labeled "no". They correspond to phases of deterministic computation in which no queries are made; the left branch represents the course of the computation on a positive answer, and the right branch to the negative answer. When no more queries are made, and the machine halts, the final configurations correspond to leaves of the tree; some of them may be accepting configurations.

Fixing the oracle set A for the machine is equivalent to selecting a path through this computation tree. At a node labeled w, the path branches left if w is in A, and it branches right otherwise. The input x is in $L(M, A)$ if and only if the path selected in the tree of M on x by the oracle A ends in an accepting leaf.

The fact that both the computations of deterministic oracle machines and the computations of nondeterministic machines are modeled by trees may induce in the reader some confusion. It is intuitive that given a nondeterministic machine, an oracle should be able, in some sense, to "guide" the nondeterministic computation through the tree to the accepting final configuration (if it exists). This can be formalized, and we believe that presenting this formalization will clarify the two notions and the differences between them. Thus, let us see how to simulate the computation of a nondeterministic machine M on input x, by a deterministic oracle machine M_o.

Define the (uncomputable) oracle set A which consists of all partial computations which are correct for M, and furthermore can be extended to a complete accepting computation. Using this oracle, the machine M_o simulates M using A to guide the computation. At each nondeterministic step, M_o queries A for information about the two possible continuations. It is described in Figure 1.4.

During the execution of this procedure, a new configuration is selected at each cycle of the main loop. This configuration is selected in such a way that it leads to an accepting configuration if one exists. Therefore, this machine accepts the same set as the nondeterministic machine M. We simulate in this way the nondeterminism by oracle queries, and construct the full accepting computation of M.

Of course, the next question that can be asked is whether the converse holds. Is it possible to simulate in a nondeterministic manner an oracle machine? An immediate negative answer follows from the fact that oracles are a much more powerful mechanism than suggested by the above simulation. For example, we could have $L(M)$ itself as an oracle, and program the oracle machine just to copy the input into the query tape, query the oracle, and accept if and only if the answer is YES. Notice that this allows to decide *any* set, given itself as oracle.

It should be observed, however, that for some restricted cases of oracle machines the converse simulation can be performed. Consider the following procedure: given an oracle machine M_o, whose oracle is not yet fixed, and an input string x, simulate step by step the behavior of M_o in a deterministic way, until M_o enters a QUERY state. Then select in a nondeterministic manner the move which could correspond to the answer of the oracle and resume the deterministic simulation of M_o.

Of course, this procedure cannot be properly called simulation, since it is not guaranteed to accept any of the sets $L(M_o, A)$. Intuitively, this procedure accepts the union of all the sets $L(M_o, A)$ where A ranges over all possible oracle sets. This follows from the fact that a nondeterministic machine accepts a word if and only if there exists an accepting computation, which

```
input x
s := the word describing the initial configuration of M on input x
comment: s keeps the current configuration
c := λ
comment: c keeps the computation up to now
loop
     if s is a final accepting configuration       .
         then halt in accepting state
     else
             append s to c as its last configuration
             obtain the set R of configurations that may legally
                 follow s obeying the transition function of M
             for each member t of R do
                 query A about the result of appending t to c
                 if the answer is YES then
                     s := t
                     exit the for loop
             end for          ʼ
             if no positive answer was obtained
                 then halt in non-accepting state
end loop
end
```

Figure 1.4 An oracle machine

corresponds to some sequence of answers given by the oracle. Moreover, this intuitive argument is not completely correct, since it requires that the oracle machine never query the same word twice in a computation; otherwise, the nondeterministic machine might guess different (hence, inconsistent) answers for different queries about the same word.

Anyway, certain restrictions can be imposed on oracle machines in order to make the simulation outlined above work properly. These restrictions consist in imposing the condition that the oracle machine always accept the same set independently of the oracle. This concept makes clearer the relationship between oracle machines and nondeterministic machines.

Definition 1.50 *An oracle machine M is robust if for every oracle A, M with oracle A halts for all inputs, and $L(M, A) = L(M, \emptyset)$.*

Example 1.51 In Figure 1.5 we present an example of a robust machine accepting the set SAT of satisfiable boolean formulas.

---

input F
check that F is syntactically correct
copy F into F'
loop
    select a variable x in F'
    if $F'|_{x:=0}$ is in the oracle then $F' := F'|_{x:=0}$
    else if $F'|_{x:=1}$ is in the oracle then $F' := F'|_{x:=1}$
    else exit the loop
until all variables are exhausted
simplify F'
if the result is "true" then accept
else apply to F the algorithm of Example 1.46
end

Figure 1.5 A robust algorithm for SAT

It is easily seen that this algorithm accepts exactly SAT, no matter how the oracle answers. This is so since it accepts either when a satisfying assignment has been found with the help of the oracle, or when the algorithm of Example 1.46 (which accepts exactly SAT) accepts. If this machine is given the same set SAT as oracle, it will accept satisfiable formulas at the end of the loop, and does not need to check all possible satisfying assignments.

Now assume that M_o is a robust oracle machine which never queries twice the same word in the same computation. Consider the nondeterministic machine M which performs the "kind of simulation" outlined above. Since M_o is robust, this implies that if $x \in L(M_o, A)$ for any A then some of (and in fact all) the halting computations of M accept x, and that $x \notin L(M_o, A)$ implies that no computation of M accept x. Therefore $L(M) = L(M_o, A)$.

Robust machines are used for discriminating between the oracles that "help" to solve a problem and those that do not. An oracle helps to solve a problem if a robust oracle machine solving that problem runs faster when using this oracle. References in which results are proved about robust machines and helping are cited in the bibliographic remarks. In this sense, SAT helps the machine of Example 1.51.

As a final remark closing this section, notice that oracle machines may as well be themselves nondeterministic. Then the computation tree has two kinds of nodes—the nondeterministic nodes and the query nodes. Fixing an oracle is equivalent to selecting the path to follow in the query nodes, and the machine accepts if and only if the resulting subtree has at least one path ending in an accepting leaf. The interaction of queries and nondeterminism allows a great increase of the computational power, which will be subject of a detailed study in Volume II.

1.9 Bibliographical Remarks

A great part of the material presented in this chapter can be found in the existing known textbooks in Theoretical Computer Science, among others Aho, Hopcroft, and Ullman (1974), Harrison (1978), Paul (1978), Lewis and Papadimitriou (1981), Hopcroft and Ullman (1979), Davis and Weyuker (1983), and Wagner and Wechsung (1986). Example 1.27 is from Gill (1977). The concept of finite automaton is usually attributed to McCulloch and Pitts (1943). The formalism used and the concept of nondeterminism are due to Rabin and Scott (1959).

In the decade of the 30's and as a direct consequence of Gödel (1931), several formalizations appear of the concept of "effective procedure". Gödel himself introduced in this work the concept of primitive recursive function as an equivalent to computation. Later, Church (1933 and 1936) introduced the λ-calculus and used it as a new formalization of the concept of procedure. Using this work, Kleene (1936) defined the concept of partial recursive function, showed it to be different from the concept of primitive recursive function, and proved the equivalence of partial recursive functions and λ-calculus. Meanwhile, a new formalization of procedure was given in Turing (1936 and 1937), which gave rise to the Turing machine formalism.

In this work, Turing also proved the equivalence of his formalism with that of Church. It is fair to say that, in a totally independent way, the Turing machine formalism was also proposed by an at the time unknown high school teacher in New York, E. Post (1936). Turing's manuscript was sent for publication a few months before Post's manuscript. In fact, Turing machines as we use them in this book are closer to Post's definition. For a nice survey of the model of Post, we recommend Uspenski (1979).

The term Church's Thesis, to denote the principle that the Turing machine and its equivalent models formalize the intuitive concept of algorithm, was coined in Kleene (1952). The Invariance Thesis has been proposed in Slot and van Emde Boas (1985).

Oracle machines arise from an idea already put forward by Turing (1936). The notion of robust machines is taken from Schöning (1985b), where the concept of helping robust machines is developed. Example 1.51 is a particular case of a theorem on sets which help themselves, which is due to Ko (1987).

2   Time and Space Bounded Computations

2.1 Introduction

We present here, in a survey style, the basic concepts and theorems about complexity classes. These classes are defined by imposing bounds on the time and the space used by the different models of Turing machines. Remember once more that the aim of this book is the study of structural properties of sets and classes of sets, and that these first two chapters are included for the sake of completeness. Readers interested in pursuing the specific topics treated in these two chapters are referred to the books indicated in the bibliographical remarks.

This chapter starts with some properties of the families of functions used to measure the rate of growth of functions. Then the definitions of running time and work space for Turing machines are defined, and some results are presented.

The functions used to bound the resources must fulfill some conditions to allow certain results to be proved. Thus, a section is devoted to presenting these conditions (namely, time and space constructibility) before presenting the main relationships between complexity classes in the last section of the chapter.

2.2 Orders of Magnitude

Consider an algorithm for solving a problem on instances whose size may be arbitrarily large. Assume that we want to analyze it in terms of its complexity, i.e. the quantity of resources required to run it. We do this by imposing bounds on the resources, and observing whether the algorithm exceeds these bounds.

However, solving the problem on longer instances can be "more difficult" than solving it on smaller ones. Therefore, it is fair to allow the amount of resources to grow with the size of the problem to be solved. Thus, in this work, the complexity of a machine will be measured by evaluating the function that, for each n, gives the quantity of resources required for instances of length n.

Some results to be proved in the next section show that the complexity of the problems is quite independent of such variations as additive constants and constant factors. Hence, our evaluation of the complexity of an algorithm will rely on the "rate of growth" of the corresponding function. In this section we define some notation to deal with this concept: the *order of magnitude* symbols O, o, Ω_∞, ω, and Θ.

Unless otherwise specified, all functions are from $\mathbb{N}-\{0\}$ into $\mathbb{N}-\{0\}$, although the definitions make sense also for real-valued functions; therefore the functions are prevented from giving the value 0. Here r denotes a real positive constant.

Definition 2.1

1. $O(f)$ *is the set of functions g such that for some $r > 0$ and for all but finitely many n, $g(n) < r \cdot f(n)$.*
2. $o(f)$ *is the set of functions g such that for every $r > 0$ and for all but finitely many n, $g(n) < r \cdot f(n)$.*
3. $\Omega_\infty(f)$ *is the set of functions g such that for some $r > 0$ and for infinitely many n, $g(n) > r \cdot f(n)$.*
4. $\omega(f)$ *is the set of functions g such that for every $r > 0$ and for infinitely many n, $g(n) > r \cdot f(n)$.*
5. $\Theta(f)$ *is the set of functions g such that for some positive constants r_1 and r_2, and for all but finitely many n, $r_1 \cdot f(n) \leq g(n) \leq r_2 \cdot f(n)$.*

There are alternative ways to define these sets of functions. For example, the notation $o(f)$ can be defined as follows:

Proposition 2.2   *The set $o(f)$ is the set of functions g such that*

$$\lim_{n \to \infty} \frac{g(n)}{f(n)} = 0$$

The proof is reduced to straightforward manipulation, and is omitted. Similar characterizations of the other orders of magnitude in terms of inferior limits and superior limits exist. This property is very useful for showing that $g \in o(f)$ for particular functions f and g, as is shown later.

Another simple property to keep in mind is the following:

Proposition 2.3 $f \in \Theta(g)$ *if and only if $f \in O(g)$ and $g \in O(f)$.*

This notation is useful for other arguments and definitions beside those regarding resource bounds. For example, recalling the concept of "subexponential density" from Chapter 1, we have the following property:

Proposition 2.4   *A set A has subexponential density if and only if its census function C_A verifies that, for each polynomial p, $C_A(p(n)) \in o(2^n)$.*

Let us present some more examples of this notation, which will be used later on.

Example 2.5 We define the function $\log^* n$, which will be used in later chapters. All our logarithms here are in base 2. Define inductively:

1. $\log^{(0)} n = n$,
2. $\log^{(k)} n = \log(\log^{(k-1)} n)$ for $k \geq 1$.

Then, $\log^* n = \min\{i \mid \log^{(i)} n \leq 1\}$.

This function grows very very slowly. The first few values of $\log^*$ are given in the table in Figure 2.1.

n	$\log^* n$
1	0
2	1
3, 4	2
from 5 to 16	3
from 17 to 65536	4
$\cdots$	$\cdots$

Figure 2.1 Rate of growth of the function $\log^*$

Using Proposition 2.2, it is easily seen that the following holds:

$$n \in o(n \cdot \log^* n), \text{ and } n/(\log^* n) \in o(n)$$

The invariance under constant factors of the sets of functions defined by the "order of magnitude" symbols provides a certain comfort in their use. For example, since the "logarithm" function only differs by a constant factor when changing the base, the use of the notation $O(\log n)$ obviates the need to specify the base. More motivation will be given by the tape compression and the linear speed-up theorems in the next section.

We can generalize these notations to families of functions as follows.

Definition 2.6 *Given a family of functions F, we denote:*

1. $O(F) = \bigcup_{f \in F} O(f)$
2. $o(F) = \bigcap_{f \in F} o(f)$
3. $\Omega_\infty(F) = \bigcup_{f \in F} \Omega_\infty(f)$
4. $\omega(F) = \bigcap_{f \in F} \omega(f)$
5. $\Theta(F) = \bigcup_{f \in F} \Theta(f)$

We state now some simple facts about orders of magnitude.

Proposition 2.7 *For each two functions f and g, f is either in $O(g)$ or in $\omega(g)$ but not in both, and f is either in $\Omega_\infty(g)$ or in $o(g)$ but not in both.*

The same property holds for families of functions:

Proposition 2.8 *For every family of functions F, and for every function f, f is either in $O(F)$ or in $\omega(F)$ but not in both, and f is either in $\Omega_\infty(F)$ or in $o(F)$ but not in both.*

We show next an interesting relationship between the Θ notation and the other notations: the Θ operator can be "absorbed" by the other operators (see the formalization of this fact below). This may seem counterintuitive at first, because it says, for example, that $\log n \in \Theta(O(n))$, while it is easily shown that $\log n$ is not in $\Theta(n)$. However, after some thought, the reader will convince himself that this fact says just that the orders of magnitude are invariant modulo "approximate" constant factors.

Proposition 2.9 *For any family of functions F,*

(a) $\Theta(O(F)) = O(F)$
(b) $\Theta(o(F)) = o(F)$
(c) $\Theta(\Omega_\infty(F)) = \Omega_\infty(F)$
(d) $\Theta(\omega(F)) = \omega(F)$

To finish, we give an example of the use of this notation. We state the simple relation between a function f and the integer approximation of the logarithm of f.

Proposition 2.10 *For every function f, $f \in \Theta(2^{\lceil \log_2 f \rceil})$.*

Proof. Define the function $g(n) = \lceil \log_2 f(n) \rceil$. It is immediately seen that for every n, $(1/2) \cdot 2^{g(n)} \le f(n) \le 2^{g(n)}$ and therefore $f(n) \in \Theta(2^{g(n)})$. □

2.3 Running Time and Work Space of Turing Machines

In order to impose bounds on the running time or on the working space of Turing machines, we must define formally these concepts.

Definition 2.11
1. *Given a deterministic Turing machine M and an input w, the computation time (or the length of the computation) of M on w is the number of computation steps required by M to halt on input w, if M halts on w; otherwise we say that the computation time of M on w is undefined.*

2. *Given a deterministic Turing machine M, its running time, sometimes denoted* $\text{time}_M(n)$, *is the (partial) function from* $\mathbb{N}$ *to* $\mathbb{N}$ *which evaluates as follows:* $\text{time}_M(n)$ *is the maximum computation time of M on an input of length n, if M halts on all of them; otherwise it is undefined.*

3. *Given a nondeterministic Turing machine M and an input w, the computation time of M on w is the number of computation steps in the shortest accepting computation of M on input w, if some computation of M accepts w; otherwise it is defined to be 1.*

4. *Given a nondeterministic Turing machine M, its running time is the function from* $\mathbb{N}$ *to* $\mathbb{N}$ *which for each n evaluates to the maximum computation time of M on an input of length n.*

5. *Given a deterministic Turing machine M and an input w, the computation space of M on w is the maximum number of tape cells scanned during the (finite or infinite) computation of M on w. If infinitely many cells are scanned in an infinite computation then we say that the computation space of M on w is undefined.*

6. *Given a deterministic Turing machine M, its work space is the (partial) function from* $\mathbb{N}$ *to* $\mathbb{N}$ *which, for each n, evaluates to the maximum computation space of M on an input of length n, if it is defined for all of them; otherwise it is undefined.*

7. *Given a nondeterministic Turing machine M and an input w, the space required by a given computation is the maximum number of tape cells scanned in the computation. The computation space of M on w is the minimum space required by an accepting computation of M on w if M accepts w, and is 1 otherwise.*

8. *Given a nondeterministic Turing machine M, its work space is the function from* $\mathbb{N}$ *to* $\mathbb{N}$ *which for each n evaluates to the maximum computation space of M on an input of length n.*

9. *We say that a language is accepted by a (deterministic or nondeterministic) machine M within time (respectively within space)* $f(n)$ *if and only if the running time of M (respectively work space of M) is at most* $f(n)$ *for every n.*

Notice that, for nondeterministic machines, the only computations which matter for measuring resources are the accepting ones. This is the reason that the time and space of non-accepting computations is 1. This decision yields easier formulations of the remaining concepts, although it violates the axioms of Blum. The formalist reader may substitute an "undefined" for the 1, and then reformulate the concept of "bound" so that only the accepting computations matter, obtaining an equivalent definition.

In some references, these definitions are given in a different manner for the nondeterministic machines, measuring the resources on *each* accepting or non-accepting computation. Although the two definitions may give rise to different classes, they will be equivalent for the time constructible and the space constructible bounds to be defined in the next section, and therefore they will be equivalent for all purposes of this book.

Recall that we do not consider functions taking value 0. Therefore, any space bound $s(n)$ to appear in this book should be considered a shorthand of the function $\max\{1, s(n)\}$. Moreover, we will not consider computations in which only a part of the input is scanned. Hence, any time bound $t(n)$ to appear in this book should be considered as a shorthand of the function $\max\{n + 1, t(n)\}$. Notice that $n + 1$ is the number of steps required to completely read the input and detect its end.

For the case of oracle Turing machines, running time is defined exactly as in the case of non-oracle machines. However, there is some controversy on the appropriate conventions for measuring the amount of space used by oracle Turing machines: should the oracle tape be counted as a work tape? We take the position that the oracle tape is to be bounded by any space bound we consider, unless otherwise indicated. We should issue the warning, anyway, that in many cases (such as, for instance, if space bounds are sublinear) we will impose on the oracle tape a space bound different from the work tape space bound.

Changing the tape alphabet is a powerful technique which allows one to show that resources can be "saved up" by a constant factor. We begin with the easiest construction, the tape compression, and later we show how to refine it to achieve a speed-up of the computation. A very similar partial speed-up will be used in the next section to characterize time constructible functions.

The *tape compression theorem* can be stated as follows:

Theorem 2.12 *If A is accepted by a Turing machine M within space $s(n)$, and c is a real number with $0 < c < 1$, then A is accepted by a Turing machine M_r within space $c \cdot s(n)$.*

Proof. Let r be an integer such that $r \cdot c \geq 2$, and assume that M has k tapes. We construct a Turing machine M_r with k tapes which performs the simulation.

Each symbol of the tape alphabet of M_r is an r-tuple of symbols of the alphabet of M. The states of M_r are tuples $(q, i_1, \ldots, i_k)$, where q is a state of M and $1 \leq i_m \leq r$. Each configuration of M_r corresponds to a configuration of M as follows: the jth symbol of each tape of M_r encodes the symbols in cells $(j - 1) \cdot r + 1$ to $j \cdot r$ of the corresponding tape of M. For each tape m, $1 \leq m \leq k$, when M's head is scanning the $(j - 1) \cdot r + l$ symbol, the corresponding head of M_r scans the j symbol, and furthermore M_r is in a state in which $i_m = l$. In this way, in each configuration, M_r knows the symbol seen by each of the heads of M, and is able to simulate it by a transition function that correctly updates the new contents of the tape and all the new components of the state.

It just remains to add to M_r a "preprocessing" to compress the input, translating it into the new alphabet. This can be done by writing down a symbol upon reading each r consecutive symbols of the input, which requires no additional space. The construction of the transition table of M_r from that of M is straightforward. The space required by M_r is $\lceil s(n)/r \rceil$. To complete the proof, it suffices to observe that r was chosen such that $r \cdot c \geq 2$. □

Notice that if the machine M is nondeterministic then M_r can also be taken to be nondeterministic, and the full proof goes through. Therefore, we can state the following:

Theorem 2.13 *If A is accepted by a nondeterministic Turing machine M within space $s(n)$, and c is a real number with $0 < c < 1$, then A is accepted by a nondeterministic Turing machine M_r within space $c \cdot s(n)$.*

A closer examination of the previous proof indicates that there is no need to simulate each step of the machine in one step. Several steps can be simulated at once, provided that the tape head does not go too far during these steps. For example, if the simulated machine stays operating for t steps within a tape region which is represented by just one tape symbol in the simulation, we can design the transition function of the simulating machine to print over the cell, in just one step, the symbol corresponding to the result of all t steps. We use this idea to prove the following lemma, which is part of the proof of the linear speed-up theorem. Our proof is "almost" the classical one, and is designed with the additional goal of simplifying the proof of Theorem 2.20 below.

Lemma 2.14 *Let M be a Turing machine having $k > 1$ tapes, and let $r > 0$ be any integer. There is a machine M_r such that, on every input w of length n, M accepts w in t steps if and only if M_r accepts w in $n + \lceil n/r \rceil + 6 \cdot \lceil t/r \rceil$ steps.*

Proof. Construct a Turing machine M_r with k tapes in a similar manner as in the previous theorem, such that each symbol of the tape alphabet of M_r is an r-tuple (which we shall call *block*) of symbols of the alphabet of M, and some of the states of M_r encode tuples $(q, i_1, \dots, i_k)$, where q is a state of M and $1 \leq i_m \leq r$. Here a difference arises with the tape compression theorem, since M_r needs other types of states to perform other computations, as described below. Configurations of M correspond to configurations of M_r as in the tape compression theorem.

The steps of M_r will be collected into groups of 6 steps. Each group of 6 steps is called a *dance*. M_r simulates exactly r steps of M in each dance. We now describe a dance. The following moves are performed simultaneously on each tape.

In the first phase, M_r spends 4 steps in recording into its finite control the contents of the cell scanned by the tape head, the contents of the cell at the left of the cell scanned by the tape head, and the contents of the cell at the right of the cell scanned by the tape head. This can be done by moving one step left, two steps right, and back one step left. Notice that, in this sequence of moves, M_r has gathered information about $3 \cdot r$ tape symbols of M.

In the second phase, M_r finds in one step the contents of these $3 \cdot r$ tape cells of M after r moves. This is a finite amount of information which can be furnished by the transition function of M_r (in some sense, this amounts to implementing those r steps in hardware!). Further, this step can be performed simultaneously with the fourth of the steps of the first phase. Observe that in r steps, the head of M can change symbols in two adjacent blocks taken from the scanned block, the block at its left, or the block at its right. It cannot write in all three (because they are r symbols long), nor can it write in other blocks either. Moreover, the head can reach only positions within one of these three blocks. Thus, the transition function of M indicates a new contents to be printed in at most two of the three blocks, and the new place for the tape head, which is in one of these two blocks.

In the third phase, the machine actually updates the contents of the tape in two moves, one to the left and one to the right. These moves must be done in the appropriate order to update the blocks. If only the middle block is to be updated, then one move suffices. In this case the machine performs an extra useless move to complete the dance.

In this way, each dance simulates r moves of M. As in the previous theorem, a preprocessing of the input is required to compress it. This can be done in n steps by reading the input, and writing in a work tape one symbol for each r symbols that are read from the input (notice that the assumption that at least two tapes are available is used here, since in one tape this compression would require more than linear time). The total computation time on an input w of length n is:

1. n steps to compress the input;
2. $\lceil n/r \rceil$ to reset the tape head to the left of the compressed input; and
3. $\lceil t/r \rceil$ dances to simulate the t computation steps.

As each dance requires 6 steps, we obtain a total of $n + \lceil n/r \rceil + 6 \cdot \lceil t/r \rceil$ steps. □

Observe that this quantity is not an upper bound on the running time of M_r, but an exact estimate. This fact will be used later on. Let us now show the *linear speed-up theorem*, which is now very easy since the basic construction has been presented in the lemma.

Theorem 2.15 *Let A be accepted by a Turing machine M with $k > 1$ tapes within time $t(n)$, and let $c > 0$ be any real number. Assume that $n \in o(t(n))$. Then A is accepted by a Turing machine M' within time $c \cdot t(n)$.*

Proof. Let r be an integer such that $r > (12/c)$, and construct the machine M_r as indicated in the previous lemma. We obtain that each word of length n in A is accepted by M_r in $n + \lceil n/r \rceil + 6 \cdot \lceil t(n)/r \rceil$ steps. For $r \geq 2$, straightforward calculation bounds above this quantity by $2n + 6 + (6/r) \cdot t(n)$. Since $n \in o(t(n))$, we obtain that for all but finitely many n this is less than $(c/2) \cdot t(n) + (6/r) \cdot t(n)$. The choice of $r > (12/c)$ guarantees a running time bounded above by $c \cdot t(n)$ for all but finitely many n. To complete the proof, it suffices to modify M_r to take care of the finitely many exceptions by table look-up. □

Again if M is nondeterministic, it can be simulated in the same manner by a nondeterministic machine M_r. Therefore we obtain:

Theorem 2.16 *Let A be accepted by a nondeterministic Turing machine M with $k > 1$ tapes within time $t(n)$, and let $c > 0$ be any real number. Assume that $n \in o(t(n))$. Then A is accepted by a nondeterministic Turing machine M' within time $c \cdot t(n)$.*

The tape compression and the linear speed-up theorems indicate that the $O(\ )$ order of magnitude notation is highly appropriate to denoting time and space bounds. Indeed, they show that given a machine M operating in time $O(t(n))$ (respectively, space $O(s(n))$), there is another machine simulating M in time $t(n)$ (respectively, space $s(n)$). Hence, we shall identify the idea of machines working in time or space f with that of machines working in time or space $O(f)$.

Other results can be proven which indicate the slowdown introduced in Turing machine computations by reducing the number of tapes. We present next such a theorem. The proof is omitted; it can be found in textbooks on Automata and Formal Language Theory.

Theorem 2.17 *For every Turing machine M working in time $t(n)$, there is a Turing machine M' with two tapes, working in time $O(t(n) \cdot \log t(n))$, such that $L(M) = L(M')$.*

A fact to be used later is that the time bound $O(t(n) \cdot \log t(n))$ even allows the alphabet of M' to be fixed "a priori", and independently of M. This fact can be observed in the standard proof of this theorem.

2.4 Time and Space Constructibility

Techniques which fall out of the scope of this book show that, for some adequately constructed bounds on any reasonable resource of Turing machines, undesirable situations can arise (such as, for example, arbitrarily large "gaps" of complexity, i.e. regions of functions which are not the complexity of any recursive function). These results indicate that it is a safe decision to impose some conditions on our resource bounds, to prevent such strange behavior. Recursiveness of the resource bounds can be shown to be insufficient. Thus, our resource bounds will always be chosen from the following classes of "well-behaved" functions:

Definition 2.18

1. *A total recursive function f from $\mathbb{N}$ to $\mathbb{N}$ is said to be time constructible if and only if there is a Turing machine which on every input of length n halts in exactly $f(n)$ steps.*

2. *A total recursive function f from $\mathbb{N}$ to $\mathbb{N}$ is said to be space constructible if and only if there is a Turing machine which on every input of length n halts in a configuration in which exactly $f(n)$ tape spaces are nonblank, and no other workspace has been used during the computation.*

Notice that the condition "on input a^n" could be substituted for "on every input of length n" where a is any symbol of the input alphabet. This can be achieved by transforming the machine in such a way that whenever it reads a symbol from the input tape, it acts as if the symbol read was an a.

Of course, time and space constructible functions are always recursive. Time constructible functions are also called *running times*. In the references using this term, the space constructible functions are usually just called "constructibles". Some examples of time and space constructible functions are given below, after the presentation of a useful theorem about them (Theorem 2.20).

Indeed, time constructible functions have sometimes been defined in a different way: f is time constructible if and only if it can be computed in $O(f)$ steps. Here "computed" means that there exists a machine that on input 1^n prints out $1^{f(n)}$ in at most $c \cdot f(n)$ steps. This second definition is interesting enough to deserve a clarification of its relationship with our time constructibility concept. In fact, we will give a sufficient condition for the two definitions being equivalent. Afterwards, we discuss the use of each of them: our definition allows "clocks" to be set on machines very easily, while the characterization is an easy-to-apply tool for proving time constructibility of functions, which may be quite a burdensome task using just our definition.

The characterization requires a lemma about time constructibility. The proof uses a partial speed-up based on Lemma 2.14.

Lemma 2.19 *Assume that the function $f_1(n) + f_2(n)$ is time constructible, where the function $f_2(n)$ is also time constructible. Assume further that there is an $\varepsilon > 0$ such that for all but finitely many n, $f_1(n) \geq \varepsilon \cdot f_2(n) + (1 + \varepsilon) \cdot n$. Then $f_1(n)$ is time constructible.*

Proof. Let M_1 be a Turing machine halting at time $f_1(n) + f_2(n)$, and let M_2 be a Turing machine halting at time $f_2(n)$. Let r be an as yet unspecified integer greater than 6. We construct a machine M_r which, for all but finitely many n, spends $f_1(n)$ steps on inputs of length n. The finitely many exceptions will be handled by table look-up.

The computation of M_r is as follows:

In a first phase, M_r compresses the input into three different work tapes. In one of them it compresses r symbols of the input in each symbol; machine M_1 will be simulated on this input. In another tape, it compresses $r - 6$ symbols of the input in each symbol; machine M_2 will be simulated on this input. This second compressed word of length $\lceil n/(r-6) \rceil$ is copied simultaneously into a separate tape for later use. All this requires n steps. During these steps, M_r counts modulo $r - 6$ using the finite control, so that at the end of the compression M_r can perform $i_1 < r - 6$ useless steps; in this way, the computation time of this phase is forced to be an integer multiple of $r - 6$. (In fact, the computation time is $(r-6) \cdot \lceil n/(r-6) \rceil$.)

In a second phase, M_r simulates simultaneously the computations of M_1 and M_2, in a sped up manner, exactly as in Lemma 2.14. Specifically, M_r simulates r steps of M_1 and $r-6$ steps of M_2 in each dance of 6 steps. M_2 will be the first to halt; then M_r stops this combined simulation. The running time $f_2(n)$ of M_2 can be a non-multiple of $r - 6$; we can assume that, when M_2 halts, the transition function informs M_r about the number $i_2 < r - 6$ of remaining steps required to reach an integer multiple of $r - 6$. Therefore, $i_2 = (r-6) \cdot \lceil f_2(n)/(r-6) \rceil - f_2(n)$.

Now M_r performs i_2 useless steps. Recalling that each dance requires 6 steps, the computation time of this second phase is $6 \cdot \lceil f_2(n)/(r-6) \rceil + i_2$.

Substituting i_2 by its value given above, this evaluates to $r \cdot \lceil f_2(n)/(r-6) \rceil - f_2(n)$. This quantity is positive for all but finitely many n.

During this phase, r steps of M_1 are simulated in each of the $\lceil f_2(n)/(r-6) \rceil$ dances. Therefore, at the end of this phase $r \cdot \lceil f_2(n)/(r-6) \rceil$ steps of M_1 have been simulated.

In a third phase, M_r continues the simulation of M_1 at the same rate as before: r steps per dance. This phase lasts for $\lceil n/(r-6) \rceil + 1$ dances; this is controlled by reading, during each dance, one of the symbols of the compressed input of length $\lceil n/(r-6) \rceil$ saved in the first phase. The computation time of this phase is $6 (\cdot \lceil n/(r-6) \rceil + 1)$, and $r \cdot (\lceil n/(r-6) \rceil + 1)$ steps of M_1 are simulated.

Comparing the running time of M_1 with the number of steps of M_1 simulated so far, it is easy to see that for sufficiently large r, and for all but finitely many n, this does not exhaust the computation time of M_1. Therefore, M_r proceeds to a fourth phase during which it simulates M_1 in real time, one step per step, until it halts; finally, M_r performs $r - 6$ useless steps and halts. The computation time for this phase is the remaining steps of M_1 plus $r - 6$, which amounts to:

$$f_1(n) + f_2(n) - r \cdot \lceil f_2(n)/(r-6) \rceil - r \cdot (\lceil n/(r-6) \rceil + 1) + r - 6$$

The value of r should be selected large enough so that this last value is positive for all but finitely many n. The hypothesis must be used to show that such an r exists. Adding up the computation times of all four phases shows that most terms cancel, yielding for M_r a computation time of exactly $f_1(n)$. Therefore f_1 is time constructible. $\qquad\square$

This lemma allows us to prove the following characterization.

Theorem 2.20 *Let f be a function for which there is an $\varepsilon > 0$ such that for all but finitely many n, $f(n) \geq (1 + \varepsilon) \cdot n$. Then f is time constructible if and only if f can be computed in time $O(f)$.*

Proof. If f is time constructible, and machine M witnesses this fact, then f can be computed in time $O(f)$ by a machine acting exactly as M which additionally writes down in a separate output tape a 1 for each step performed by M.

Conversely, let M be a Turing machine that computes $f(n)$ within $O(f(n))$ steps, and let $g(n)$ be the computation time of M. We have that $g(n) \leq cf(n)$ for some constant c. We show that $f_1(n) = f(n)$ and $f_2(n) = g(n)$ satisfy the hypothesis of the previous lemma. By definition, $g(n)$ is time constructible. The time constructibility of $f(n) + g(n)$ is witnessed by a machine that uses M to compute $f(n)$ in unary (in $g(n)$ steps) and then counts the number of 1's of the result, spending $f(n)$ steps in that.

Let us check that there is an $\varepsilon > 0$ such that for all but finitely many n,

$$f_1(n) \geq \varepsilon \cdot f_2(n) + (1 + \varepsilon) \cdot n$$

as required in the lemma. Let $\varepsilon_1, \varepsilon_2, \varepsilon_3$, and ε_4 be positive real numbers such that the following holds:

1. for all but finitely many n, $f(n) \geq (1 + \varepsilon_1) \cdot n$;
2. $(1 + \varepsilon_1) \cdot (1 - \varepsilon_2) > 1$;
3. $\varepsilon_3 = (1 + \varepsilon_1) \cdot (1 - \varepsilon_2) - 1$;
4. $\varepsilon_4 = \min\{\varepsilon_2/c, \varepsilon_3\}$.

Then, for all but finitely many n,

$$f(n) = \varepsilon_2 \cdot f(n) + (1 - \varepsilon_2) \cdot f(n) \geq (\varepsilon_2/c) \cdot g(n) + (1 - \varepsilon_2) \cdot (1 + \varepsilon_1) \cdot n$$
$$= (\varepsilon_2/c) \cdot g(n) + (1 + \varepsilon_3) \cdot n \geq \varepsilon_4 \cdot g(n) + (1 + \varepsilon_4) \cdot n$$

which was to be shown. $\qquad\square$

A similar characterization can be proved even if the definition of "computed in time $O(f)$" refers to a binary input and/or output. See the bibliographic remarks. Observe also that the analogous characterization of space contructibility is immediate. It can be stated as follows:

Theorem 2.21   *A function f is space constructible if and only if f can be computed in space $O(f)$.*

We can use this characterization to prove the time constructibility of many functions used as time bounds throughout this book.

Example 2.22   The following functions are time constructible:

1. n^k for every fixed $k \geq 1$.
2. $n!$
3. $2^{c \cdot n}$ for every fixed c.
4. 2^{n^k} for every fixed $k \geq 1$.
5. $n \cdot \lceil \log n \rceil^k$ for every fixed $k \geq 1$.
6. $n \cdot \log^* n$.

The proof amounts to an immediate check that the hypothesis of Theorem 2.20 holds. Similarly, many other functions can be shown to be time constructible.

The interest of defining time constructible and space constructible functions as in Definition 2.18 is that this allows us to impose bounds on the computations of Turing machines in an effective (and easy) manner. Let us briefly discuss this point.

First of all, note that in general, given f, it is not decidable whether, for every n, a given machine halts within $f(n)$ steps on every input of length n. This problem is not even recursively enumerable. Therefore, when we define a complexity class by bounding the running time, we cannot hope to discriminate in an effective manner which machines witness the membership of a set to the class, and which ones do not. In other words, we are not able to effectively enumerate all the Turing machines which will always operate within a given time or space bound.

However, it will be shown in many places in this book that such an effective discrimination is crucial to showing most of the results about complexity classes. Thus, given a time constructible time bound, we must construct an enumeration of machines which obey the time bound and cover all the sets in the class.

This is done as follows. Let t be the time bound, and let M be the machine which witnesses its time constructibility. Start from an enumeration of all Turing machines, $M_1, M_2, \ldots$, and construct a new enumeration of machines $M_1', M_2', \ldots$, such that M_i' consists of running in parallel machine M_i and machine M. If M_i is the first to halt, M_i' accepts or rejects accordingly, but if M stops first then M_i' rejects.

In this way, the enumeration $L(M_i')$ consists exactly of all the sets decidable in time t. Indeed, $L(M_i')$ is decidable in time t since M_i' decides it in time t; and conversely, if any set A is decidable in time t, say via machine

M_i, then the simulation of M_i will be the first to halt when running M_i', and then $L(M_i') = A$.

Observe that this process requires the time bound to be time constructible. We refer to this process as *clocking* machines by the time bound, because the process may be compared with a machine having set a built-in "alarm watch" and halting when the alarm "sounds". A very similar process can be carried out with space constructible functions to ensure that a machine obeys a given space bound. The description of this procedure is left to the reader.

Time constructible functions can grow as fast as any other recursive function. This fact will be used in Chapter 7.

Lemma 2.23 *For every recursive function f, there is a recursive function g which is time constructible and such that $\forall n\, g(n) > f(n)$.*

Proof. Let M be a Turing machine which computes f in unary: on input 1^n, M outputs $1^{f(n)}$. Consider a machine M' which on input x of length n, writes 1^n on its work tape, and then simulates M on 1^n. Clearly, M' spends the same time on every input of length n.

Define the function $g(n)$ as the number of steps that M' spends on inputs of length n. Then g is time constructible, because it is the running time of M'. Just to convert x into 1^n and to write down $1^{f(n)}$ in the output tape, M' requires at least $n + f(n)$ steps, and therefore $g(n) > f(n)$.   □

2.5 Bounding Resources: Basic Definitions and Relationships

Definition 2.24 *For any functions $t(n) \geq n + 1$ and $s(n) \geq 1$, define:*

1. *$DTIME(t)$ is the class of all sets accepted by deterministic Turing machines whose running time is bounded above by $t(n)$.*
2. *$NTIME(t)$ is the class of all sets accepted by nondeterministic Turing machines whose running time is bounded above by $t(n)$.*
3. *$DSPACE(s)$ is the class of all sets accepted by deterministic Turing machines whose work space is bounded above by $s(n)$.*
4. *$NSPACE(s)$ is the class of all sets accepted by nondeterministic Turing machines whose work space is bounded above by $s(n)$.*

A similar notation is used when a family of functions is used as resource bounds: for example, $DTIME(F) = \bigcup_{f \in F} DTIME(f)$, and similarly for all the other complexity classes. Also, when oracle machines are considered, analogous definitions can be stated about complexity classes with oracles. It should be noticed, though, that a given oracle machine may spend substantially more resources on the same input if the oracle set is changed. Hence, the classes obtained depend heavily on the oracle, not only in that the accepted sets vary from one oracle to another, but also in that the running time and the work space of the machines change as well.

No special notation is introduced for the case that the oracle tape of the machines defining a given class is bounded by a function different from the work tape space bound. Whenever this is the case, a verbal statement will be provided, as in "consider machines with a work space bound f and an oracle space bound g".

Of course, time bounds imply space bounds, because in t steps no more than t cells can be scanned. In a certain sense, the converse implication also holds, and throughout this book we will use frequently (and not always explicitly) the following simple argument: if a deterministic machine repeats twice the same configuration in the course of a computation, then this computation is infinite and the machine will never halt. Indeed, the path leading from the configuration to itself will be followed again, and the configuration will repeat infinitely often; the machine falls into an infinite loop, and will never accept.

A somewhat different argument applies to nondeterministic machines. In this model it is possible that a configuration repeats and the machine still halts, because a different computation path is taken. However, even in this case we can get rid of computations in which configurations repeat, since if there is an accepting computation in which some configuration repeats, then it is enough to "shortcircuit" the loop in the computation and avoid the repetition.

Therefore, for the languages accepted by both deterministic and nondeterministic machines, a space bound implies a time bound, in the following sense: assume that the language A is accepted by a space bounded machine. Clock the machine, as indicated in the previous section, by a time constructible function t that majorizes the number of different configurations (such a function exists by Lemma 2.23). Then the clocked machine accepts the same language A, and it is time bounded.

However, if this time constructible function t is substantially larger than the number of configurations, the new machine might use more resources than the old one. Space constructibility may be used to avoid this undesirable situation, as the following lemma indicates.

Lemma 2.25 *Let M_1 be a Turing machine which uses at most $s(n) \geq \log n$ space on every input of length n, where s is space constructible. Then there exists a Turing machine M_2 which accepts the same language as M_1 in space $s(n)$, and halts on every input.*

Proof. It is easily seen that the number of configurations using space $s(n)$ is bounded above by $2^{c \cdot s(n)}$ for some c. Thus, M_2 first lays off $s(n)$ tape cells on an extra tape, and then simulates M_1 in the remaining tapes while counting on the extra tape, in base c, the number of moves simulated so far. When the count reaches $2^{c \cdot s(n)}$, i.e. when the counter fills up the $s(n)$ squares previously laid out, M_2 knows that M_1 is looping forever and stops the simulation in a rejecting state.   □

The machine obtained in this way uses the same space as M_1, and runs for as much time as M_1 might run without looping. Therefore, no more resources can be saved under the given hypothesis.

This lemma can be strengthened by using other techniques. First, the space constructibility of $s(n)$ can be made unnecessary if $s(n) \geq \log n$, by letting the length of the counter (initially $\log n$) grow with the number of tape cells scanned by M_1. Furthermore, even this hypothesis can be relaxed by using a different technique which amounts, briefly, to "simulating M_1 backwards". (See the references at the end of the chapter.)

Some relationships that hold among the complexity classes can be formalized as follows:

Theorem 2.26   Let $s(n) \geq 1$, $s'(n) \geq 1$, $t(n) \geq n + 1$, and $t'(n) \geq n + 1$ be functions. Then:

(a) $DTIME(t) \subseteq NTIME(t)$, and $DSPACE(s) \subseteq NSPACE(s)$.
(b) $DTIME(t) \subseteq DSPACE(t)$, and $NTIME(t) \subseteq NSPACE(t)$.
(c)   If $s' \in O(s)$ then

$$DSPACE(s') \subseteq DSPACE(s) \text{ and } NSPACE(s') \subseteq NSPACE(s)$$

(d)   If $s' \in \Theta(s)$ then

$$DSPACE(s') = DSPACE(s) \text{ and } NSPACE(s') = NSPACE(s)$$

(e)   If $t' \in O(t)$ and $n \in o(t)$ then

$$DTIME(t') \subseteq DTIME(t) \text{ and } NTIME(t') \subseteq NTIME(t)$$

(f)   If $t' \in \Theta(t)$ and $n \in o(t)$ then

$$DTIME(t') = DTIME(t) \text{ and } NTIME(t') = NTIME(t)$$

(g)   If s is space constructible and $s(n) \geq \log n$ then

$$NSPACE(s) \subseteq DTIME(2^{O(s)})$$

(h)   If t is time constructible then $NTIME(t) \subseteq DSPACE(t)$.
(i)   If s is space constructible and $s(n) \geq \log n$ then

$$DSPACE(s) \subseteq DTIME(2^{O(s)})$$

(j)   *If t is time constructible then $NTIME(t) \subseteq DTIME(2^{O(t)})$.*
(k)   *If $s(n) \geq \log n$ then $NSPACE(s) = co\text{-}NSPACE(s)$.*

Proof.

(a)   Immediate since every deterministic machine can be considered as a nondeterministic machine.
(b)   In t steps, at most $t+1$ tape cells can be scanned by the tape heads.
(c)   By the definition of $O(s)$, there is a positive constant c such that $s' \leq c \cdot s$. Any set A in $DSPACE(s')$ is therefore in $DSPACE(c \cdot s)$. By the tape compression theorem (Theorem 2.12), A is in $DSPACE(s)$. The same argument proves the nondeterministic case.
(d)   Follows from (c) and Proposition 2.3.
(e)   Analogous to (b), using the linear speed-up theorem (Theorem 2.15). Note that the hypothesis $n \in o(t)$ is required to apply this theorem.
(f)   Follows from (e) and Proposition 2.3.
(g)   Let M be a nondeterministic machine using space s. Without loss of generality, we can assume that if M accepts, it erases every tape and brings the tape heads to the first cell before entering the accepting state. Thus, there is only one possible accepting configuration. By the argument above, we know that accepting computations do not repeat configurations. By the definition of a space bounded nondeterministic machine, we know also that for each input accepted there is a computation in which all the configurations use space $s(n)$. The number of configurations of M on an input x of length n, using space $s(n)$, is $O(2^{O(s)})$.

Consider a deterministic machine M' that, on input x of length n, generates a graph having all the configurations of M on input x as vertices, setting a directed edge between two configurations if and only if the second one is reachable from the first in one step according to M's transition function. The number of configurations is computed using the space constructibility of s. Compute next whether there is a directed path in the graph joining the initial and the unique accepting configuration, and accept if and only if this is the case. This can be done in time polynomial in $2^{O(s)}$, which is $O(2^{O(s)})$. M' accepts the same language as M does, and therefore this language lies in $DTIME(2^{O(s)})$.

(h)   Let M be a nondeterministic machine accepting a set A in time t. Using the time constructibility of t, simulate t steps of each possible computation path of M, until an accepting one is found; then accept. If no accepting path is found, then reject. All computations are simulated re-using the same space. To keep track of the currently simulated computation, a string of t bits suffices (to indicate whether the left or the right branch was followed at each step). Therefore the space required is t.

Parts (i) and (j) follow directly from parts (a), (b), and (g). The proof of part (k) is given in the appendix. □

Some of the constructibility conditions of this theorem can be partially relaxed. As indicated in our comment to Lemma 2.25, space constructibility may be dropped in some cases, adding counters to the machines to bound the number of steps and letting the counters grow as the computation proceeds. (See also the references in the bibliographical·remarks.) Also, it should be noticed that those parts of this theorem referring to the class $DTIME(2^{O(s)})$ are not meaningful unless $s(n) \geq \log n$. The reason is that our convention on time bounds implies that every time bound is at least linear, and hence $DTIME(2^{O(s)})$ is in fact $DTIME(\max\{2^{O(s(n))}, n\})$.

Some of the relationships stated in Theorem 2.26 do not hold under relativization, unless great care is taken that the oracle bounds are appropriate. For instance, a long query tape may blow up the number of configurations. Thus, usually we admit that oracle tapes be bounded by the same function bounding the oracle space. Exceptions to this rule will be explicitly indicated.

We state next a very interesting nontrivial relationship, based on a clever simulation of nondeterministic computations by means of a "divide and conquer" scheme. This technique will be used in other places in this book. The result is widely known as Savitch's Theorem, after its author. We present first the theorem for oracle machines.

Theorem 2.27 *Let $s(n) \geq \log n$ be a space constructible function. Let M_1 be a nondeterministic oracle Turing machine and let A be an oracle set. Assume that M_1 with oracle A works in space $s(n)$. Then there is a deterministic oracle machine M_2 such that with oracle A, it accepts the same set as M_1 and works in space $s^2(n)$.*

Proof. As in part (g) of the previous theorem, the number of different configurations of M_1 on a given input is bounded by $2^{\lceil c \cdot s(n) \rceil}$ for some constant c. Therefore if M_1 accepts then it does so with a computation consisting of at most $2^{\lceil c \cdot s(n) \rceil}$ configurations, each of a size of $O(s(n))$. Notice that the fact that $s(n) \geq \log n$ is used here, since $\log n$ space is needed just to record the position of the input tape head.

The deterministic oracle machine M_2 simulates the computation of M_1 on input x with oracle A by calling a recursive boolean function "reachable(I_1, I_2, j)", which returns the boolean value "true" if there exists a computation from I_1 to I_2 consisting of at most 2^j steps. The machine M_2 is defined in Figure 2.2, and the function "reachable" is given in Figure 2.3.

It can be observed that the key of the simulation is the function "reachable", which decides whether there is a partial computation of length at most 2^j between two configurations. It does so by looking for the middle configuration I, and checking recursively that it is indeed the middle configuration.

```
input x
let I_i be the initial configuration of M_1 on x
for each final configuration I_f of M_1 do
     if reachable(I_i, I_f, ⌈c·s(n)⌉)
          then accept and halt
reject
end
```

Figure 2.2   Deterministic simulation of nondeterministic machines

This checking amounts to verifying the existence of two partial computations of length at most 2^{j-1} each. The crucial point is that the check for the second half of the computation can be made re-using the same work space used for the first half!

It is immediately clear that M_1 accepts its input if and only if M_2 does, provided the oracle set is the same. Let us show the space bound on M_2. To simulate the recursive calls, M_2 uses a working tape as a stack, storing in it the information corresponding to successive calls of the function. Each call diminishes by 1 the value of j. Therefore the depth of the recursion is the

```
function reachable(I_1, I_2, j) returns boolean
if j = 0 then
     if I_1 = I_2 or I_2 is reached in one step from I_1
     then return true
     else return false
comment: this may require a query to A in case that I_1 is a query configuration
     end if
else for each possible configuration I of size
     less than or equal to s(n) do
comment: check whether I is the middle configuration
          if reachable(I_1, I, j − 1) and reachable(I, I_2, j − 1)
               then return true
          end if
     end for
     return false
comment: no middle configuration found, no path exists
end if
end
```

Figure 2.3   The subroutine "reachable"

initial value of j, which is $O(s(n))$. No more than $O(s(n))$ calls are active simultaneously.

For each call we must store the current values of I, I_1, and I_2, of size $O(s(n))$ each, and the current value of j, of size at most $\log s(n)$ in binary, together with some constant size bookkeeping. Therefore $O(s^2(n))$ space suffices to hold the whole stack.

Notice, finally, that the fact that $s(n)$ is space constructible is used to lay out $O(s(n))$ blocks of size $O(s(n))$ each, to hold the stack, and for controlling the size of the configuration I within each call, as well as for knowing in advance the value of $\lceil c \cdot s(n) \rceil$ to start the recursive process.   □

As a corollary, we can get a version of Savitch's theorem for machines without oracle.

Corollary 2.28 *If $s(n) \geq \log n$ is space constructible, then $NSPACE(s(n))$ is included in $DSPACE(s^2(n))$.*

A point is in order here: the fact that the query tape is bounded by the space bound as well is crucial for the proof. It can be shown that if no space bound (or a too large one) is placed on the oracle space then Theorem 2.27 fails to hold.

These last results show the relationships between bounding different resources in different models, deterministic and nondeterministic. We move to studying the relationships between classes when the same resource in the same model is bounded by different functions. A sufficient condition has been presented for the equality of the classes in parts (d) and (f) of Theorem 2.26. We now present some sufficient conditions for the strict inclusion between deterministic time and space classes. These results are known as the "space hierarchy" and the "time hierarchy" theorems. The question we address in them can be expressed as: "how much larger must be a resource bound to allow strictly more computational power?" The answer is found by a diagonalization argument.

Theorem 2.29 *Let s and s' be space constructible space bounds, and assume that $s' \in \omega(s)$. Then $DSPACE(s')$ contains a language which is not in $DSPACE(s)$.*

Proof. We construct by diagonalization a set in $DSPACE(s')$ which is not in $DSPACE(s)$. To prove that it is in $DSPACE(s')$, we define this set by means of a machine M which accepts it in space s'. The machine M is given in Figure 2.4.

It is immediate that M works in space s'. Let A be the language accepted by M. We show that A is not in $DSPACE(s)$. Assume that some machine M' with encoding u accepts A in space s, and let t be the number of tape symbols of M'. By Lemma 2.25, we can assume that M' halts on every input.

input w
let n be the length of w
using the space constructibility of s', lay out
    $s'(n)$ tape cells in the work tapes, and abort
    the computation (rejecting the input) whenever
    it attempts to leave this region
look for and skip the (unique) prefix of w of the form 1^*0
check that the remaining part of w encodes a deterministic
    Turing machine M_w and simulate it on input w
comment: each of the t tape symbols of M_w is represented by $\log t$ bits
if the simulation can be completed and ends in a rejecting state
    then accept
    else reject
end

Figure 2.4 A diagonalizing machine

Since $s' \in \omega(s)$, there exists an n greater than $|u|$ such that $\log t \cdot s(n) < s'(n)$. On the input $w = 1^j 0u$ of length n, M has enough space to complete the simulation, and w is accepted by M if and only if w is rejected by M'. Therefore M' does not accept A, contrarily to the assumption. $\qquad\square$

A stronger version of the space hierarchy can be shown, in which the space constructibility of s is unnecessary. This requires one to prove that Lemma 2.25 holds even under a weaker hypothesis. See again the references.

We proceed now to the time hierarchy theorem. The proof is via a similar diagonalization, and has as its main difference the fact that the hypotheses are rather stronger. The reason is that the diagonalizing machine has a fixed alphabet and a fixed number of tapes, but it must simulate machines with an unbounded alphabet and an unbounded number of tapes. This introduces a logarithmic slowdown.

Theorem 2.30 *Let t and t' be time bounds such that t' is time constructible and $t' \in \omega(t \cdot \log t)$. Then $DTIME(t')$ contains a language which is not in $DTIME(t)$.*

Proof. This is very similar to the previous proof. A machine M runs the same algorithm as in that proof, except for the construction and layout of the work tape; instead, it runs in parallel a machine that shuts it off after t' steps have been performed. The time constructibility of t' ensures that this is possible.

Since the machine M_w may have arbitrarily many tapes, and its alphabet may be arbitrarily large, $t \cdot \log t$ steps may be required for the simulation, as indicated by Theorem 2.17 and the comment following it. However, the stronger hypothesis that $t' \in \omega(t \cdot \log t)$ implies the existence of inputs being long enough to perform the complete simulation. Therefore the argument which shows that the resulting language is not in the class over which the diagonalization is performed carries over, showing that this language is not in $DTIME(t)$ as required.   □

Both hierarchy theorems can be generalized to broader classes. For instance, it is possible to produce sets in the class $DTIME(t)$ which are not in the class $DTIME(F)$, provided that F is a family of reasonable bounds growing sufficiently slower than f; and similarly with space. The proof is the same, and they are left to the reader.

2.6 Bibliographical Remarks

The notations for orders of magnitude were proposed in the classical work by Knuth (1976), based on several older mathematical notations. We follow Vitányi and Meertens (1984). Some of the results in that section are pointed out in Balcázar and Gabarró (1986).

The study of time and space complexity classes got its big push with Hartmanis and Stearns (1965), Stearns, Hartmanis, and Lewis (1965), and Lewis, Stearns, and Hartmanis (1965). This work contains most of the basic theorems of complexity classes, including the Tape Compression Theorem (Theorems 2.12 and 2.13), the Linear Speed-up Theorem (Theorems 2.15 and 2.16), and the Time and Space Hierarchy Theorems (Theorems 2.29 and 2.30). Most of the results in these works are included in many of the textbooks on basic Theoretical Computer Science which we mentioned in the previous chapter. However, in our opinion, some of these books present some material in a more obscure way than in the original work. An excellent presentation of this material is done by Paul (1978), but unfortunately there is no English version of this book.

A number of previous works studied similar aspects of computation; among them we shall mention Yamada (1962) and Ritchie (1963). Theorem 2.17 is from Hennie and Stearns (1966); see also Theorem 4.19 of Paul (1978). The non-relativized version of Savitch's theorem is from Savitch (1970), and the adaptation of the proof to the relativized version follows straightforwardly from it. References discussing the issue of the oracle tape bound for space bounded oracle machines are Ladner and Lynch (1976) and Ruzzo, Simon, and Tompa (1984); many other authors have reported results on this subject. We will address these questions again later in this book.

The equivalence of the two definitions of constructibility is from Kobayashi (1985). Further results about constructibility are obtained there, as for example the case in which inputs and/or outputs are given in binary, or a theorem to deal with functions not satisfying the hypothesis of Theorem 2.20.

The improvements of Lemma 2.25 mentioned in the text are as follows: the classic version of the result does not require s to be space constructible, but only $s(n) \geq \log n$. The ideas required for the proof can be found in the proof of Theorem 4.3 of Paul (1978). A more recent improvement appears in Sipser (1980), where a technique of "simulating a computation backwards" is used to show a theorem from which Lemma 2.25 follows, with no hypothesis at all on s. The same reference points out that this result allows one to remove all conditions on s in the space hierarchy theorem. However, it is known from results in axiomatic complexity theory that the space constructibility of s' cannot be removed.

3 Central Complexity Classes

3.1 Introduction

In this chapter we present some of the ground work for the following chapters. We define the basic concepts of complexity theory, and prove the basic facts about them. Most of the concepts and results can be found in some other textbooks, although several proofs are presented in a different form. A substantial part of the remaining chapters will depend heavily on the notation and the results presented in this one.

We begin by defining the basic complexity classes and their known relationships. All the classes are presented in Section 3.2 in terms of sets and accepting machines. The corresponding classes of functions and their basic properties are presented in Section 3.3. We then define one of the main tools for the study of complexity classes: the polynomial time m-reducibility, and related concepts, such as closure, completeness, and hardness.

Then follows a section presenting some well-known NP-complete problems. It is not the purpose of this book to study the algorithmic treatment of the NP-complete sets; we refer the interested reader to the excellent book by Garey and Johnson, which can be found in the references. We present next a $PSPACE$-complete problem which will play an important role in some of the remaining chapters.

The next section is devoted to a proof technique, known as "padding", which allows some nontrivial inequalities among complexity classes to be proved. To end the chapter, we show the basic facts about a stronger form of reducibility: the logarithmic space m-reducibility.

3.2 Definitions, Properties, and Examples

We begin by defining the following complexity classes:

Definition 3.1
$$LOG = \bigcup_{c \geq 1} DSPACE(c \cdot \log_2 n)$$
$$NLOG = \bigcup_{c \geq 1} NSPACE(c \cdot \log_2 n)$$
$$P = \bigcup_{i \geq 0} DTIME(n^i)$$

$$NP = \bigcup_{i \geq 0} NTIME(n^i)$$
$$PSPACE = \bigcup_{i \geq 0} DSPACE(n^i)$$
$$NPSPACE = \bigcup_{i \geq 0} NSPACE(n^i)$$
$$DEXT = \bigcup_{c \geq 0} DTIME(2^{c \cdot n})$$
$$NEXT = \bigcup_{c \geq 0} NTIME(2^{c \cdot n})$$
$$EXPSPACE = \bigcup_{c \geq 0} DSPACE(2^{c \cdot n})$$
$$EXPTIME = \bigcup_{c \geq 0} DTIME(2^{n^c})$$
$$NEXPTIME = \bigcup_{c \geq 0} NTIME(2^{n^c})$$

The inclusion relationships among these classes are presented in the following proposition.

Proposition 3.2   *The following properties hold:*

(a)   *Each deterministic class is closed under complementation.*
(b)   *Each deterministic class is included in its nondeterministic counterpart.*
(c)   *NLOG is included in P.*
(d)   *PSPACE = NPSPACE.*
(e)   *NP is included in PSPACE.*
(f)   *PSPACE is included in EXPTIME.*
(g)   *NLOG is strictly included in PSPACE, which in turn is strictly included in EXPSPACE.*
(h)   *P is strictly included in DEXT, which in turn is strictly included in EXPTIME.*
(i)   *NLOG is closed under complementation.*

Proof.

(a)   Exchange accepting and rejecting final states in the deterministic Turing machines that define the class.
(b)   By Theorem 2.26(a).
(c)   By Theorem 2.26(g).
(d)   By Corollary 2.28, for each polynomial n^i, $NSPACE(n^i)$ is included in $DSPACE(n^{2 \cdot i})$. Thus, the union of all the classes $NSPACE(n^i)$ is included in the union of the classes $DSPACE(n^i)$.
(e)   By Theorem 2.26(h).
(f)   By Theorem 2.26(g).
(g)   By Corollary 2.28, $NLOG$ is included in $DSPACE(\log^2 n)$, which by Theorem 2.29 is strictly included in $PSPACE$, which in turn is strictly included in $EXPSPACE$ by an analogous argument.
(h)   Follows from the same argument as in Theorem 2.30.
(i)   By Theorem 2.26(k).   □

At the moment, no other strict inclusions or equalities are known. Thus, all the remaining relationships are open problems. These open questions present two of the most pressing global problems in the field of structural complexity: the trade-off between determinism versus nondeterminism, exemplified by the problem $P \stackrel{?}{=} NP$, which in a certain way was the seed for the theory; and the trade-off between resources, exemplified by the problem $P \stackrel{?}{=} PSPACE$. A great deal of this book is devoted to clarifying the relationships between the above classes, and to gaining as many intuitions as possible about the characteristics and properties of these classes.

The most interesting class from the point of view of this book is surely the class NP, the class of problems which can be solved in polynomial time using a nondeterministic procedure. Such interest lies in the fact that this class contains many practical problems that can be identified by the following property: *there is no known way to compute a solution in polynomial time, but there is a known way to check in polynomial time whether a potential solution is an actual solution.*

Thus, our final goal could be expressed as follows: if the problems in NP can be solved in polynomial time (i.e. $P = NP$), *what is the way* to do it? And if the problems in NP cannot be solved in polynomial time (i.e. $P \neq NP$), *what is the reason* for this difficulty? Finally, in either of the two cases, *why is it so difficult* to answer these questions?

Let us present some examples of sets in NP. The first example is the satisfiability problem, which was already defined in Chapter 1. Recall that the satisfiability problem, SAT for short, is to determine, given a boolean formula, whether it is satisfiable.

Using the nondeterministic algorithm presented in Figure 1.3, we can establish the following theorem:

Theorem 3.3 SAT $\in NP$.

It may seem that the fact of proving that a given problem is in NP is an easy task. The next problem which we will present indicates that this consideration is wrong. Recall that the set of positive integers has a canonical encoding in binary over the alphabet $\{0, 1\}$. Let PRIMES be the language formed by all the words encoding prime numbers, and let COMPOSITES be the language formed by all the words encoding non-prime numbers. As we show next, it turns out that both sets are in NP.

Theorem 3.4
(a) *COMPOSITES* $\in NP$.
(b) *PRIMES* $\in NP$.

Proof.

(a) In Figure 3.1, a nondeterministic algorithm that tests in polynomial time whether a number n belongs to COMPOSITES is presented.

```
input n
nondeterministically guess numbers n₁ and n₂ less than n
if n = n₁·n₂ then accept
end
```

Figure 3.1 A nondeterministic algorithm for COMPOSITES

(b) We will make use of Fermat's Theorem, which states that any integer $n > 2$ is a prime if and only if there exists an integer x, with $1 < x < n$, such that

(a) $x^{n-1} \equiv 1 \pmod{n}$, and
(b) For every i with $1 \le i < n - 1$, $x^i \not\equiv 1 \pmod{n}$.

To test whether a given number is a prime, consider the algorithm of Figure 3.2. Observe that if it holds that $x^{n-1} \equiv 1 \pmod{n}$, then the least i such that $x^i \equiv 1 \pmod{n}$ must divide $n - 1$; moreover, any multiple of i, say $a \cdot i$, must also satisfy $x^{a \cdot i} \equiv 1 \pmod{n}$. Therefore if there is an i such that $x^i \equiv 1 \pmod{n}$ then there exists a p_j such that $x^{(n-1)/p_j} \equiv 1 \pmod{n}$.

It can be seen that the procedure uses a number of steps bounded by $O((\log n)^5)$. Observe that $\log n$ is the length of the input n.   □

```
input n
if n = 2, then accept
if n = 1 or n is an even integer greater than 2, then reject
if n is odd and greater than 2
     then guess an x with 1 < x < n, and verify that
     (i)   xⁿ⁻¹ ≡ 1  (mod n)
     (ii)  to verify the second condition in Fermat's theorem:
              guess a prime factorization p₁...pₖ of n − 1
              recursively check that each pⱼ is a prime
              check that n − 1 = p₁ · p₂ ··· pₖ
              check that xⁿ⁻¹/ᵖⱼ ≢ 1  (mod n)
          if all these conditions hold then accept
end
```

Figure 3.2 A nondeterministic algorithm for PRIMES

For a long time, many researchers have been looking for a deterministic algorithm which could solve the problem in polynomial time, but so far no such algorithm has been proved correct. Note that the easy algorithm which consists of checking all the potential divisors is at least linear in the input number, hence exponential in the length of the input, which is coded in binary. The membership of PRIMES to P is one of the challenging problems which are open in Computer Science. We shall return to this problem in Chapter 6, where we study randomized algorithms.

3.3 Computing Functions: Invertibility and Honesty

Recall from Chapter 1 that Turing machines can be used to compute functions, by furnishing them with an output tape. A Turing machine computes a partial function f if and only if it accepts the domain of f and the accepting computation writes down the image of the input under f.

Of course, resource bounds may be imposed on the machines computing functions as well as on accepting machines. Thus, some functions can be computed within time or space bounds, and we obtain complexity classes of partial functions. In particular, we are interested in functions that are "easy to compute", in the sense that they can be computed in time bounded by a polynomial in the length of the input.

We will consider here some issues regarding the possibility of computing inverses of "easy to compute" functions. An *inverse* of a partial function f is any partial function g such that, for every y in the range of f, $f(g(y)) = y$.

We define the following complexity classes of functions.

Definition 3.5 *PF is the class of (partial) functions that can be computed in polynomial time.*

Definition 3.6 *PSPACEF is the class of (partial) functions that can be computed in polynomial space.*

The following simple result gives a characterization of P and of *PSPACE*, by showing that it is equivalent to consider sets or characteristic functions.

Proposition 3.7 *A set is in P (PSPACE, respectively) if and only if its characteristic function is in PF (PSPACEF, respectively).*

Proof. If a set A is in P, then its characteristic function can be computed in polynomial time by running the machine that accepts A in polynomial time and writing down as output 1 or 0 in accordance with the result of the machine.

Conversely, if the characteristic function of a set is computable in polynomial time then we can decide the set by computing this function and accepting if and only if the output is 1.

An identical argument works for *PSPACE*.   □

Corollary 3.8 *If PSPACEF = PF, then PSPACE = P.*

It is interesting to note that the reverse implication also holds.

Theorem 3.9 *If P = PSPACE then PF = PSPACEF.*

Proof. Assume that $P = PSPACE$, and let f be any function computable in polynomial space. Define the following set:

$$\text{prefix}(f) = \{\langle x, y \rangle \mid \text{there is a } z \text{ such that } f(x) = yz\}$$

Then prefix(f) is in *PSPACE*, since we can search for z using an amount of space bounded by $|f(x)|$. By hypothesis, prefix(f) is in P. The algorithm given in Figure 3.3 computes $f(x)$. This proof method is quite useful, and is known as "prefix searching".

```
input x
y := λ
loop
     if ⟨x, y0⟩ ∈ prefix(f) then y := y0
     else if ⟨x, y1⟩ ∈ prefix(f) then y := y1
     else exit
end loop
output y
end
```

Figure 3.3 Computing $f(x)$ by "prefix searching"

The outer loop is performed $|f(x)|$ times, which is polynomial on $|x|$. Each loop needs to decide membership in the set prefix(f), which is in P, of a string of length polynomial in $|x|$. Hence, the running time is polynomial on $|x|$. Thus f is computable in polynomial time.   □

We consider now the possibility of computing an inverse of a partial function. By mapping long inputs to small outputs, it is easy to construct a function whose inverse is not a *PF* function. We discuss this fact in the remaining part of the section. First we need a definition:

Definition 3.10 *A function f is honest if and only if for every value y in the range of f there is an x in the domain of f such that $f(x) = y$ and $|x| \leq p(|y|)$ for some fixed polynomial p.*

An alternative way to define "honesty" is presented in Exercise 3. We are going to present some observations relating the honesty of f to the possibility of computing an inverse of f.

Proposition 3.11 *If f has an inverse computable in polynomial time, then f is honest.*

Proof. Let g be an inverse of f computable in polynomial time, and let p be a polynomial bounding the time needed to compute g. As the machine computing this inverse is able only to perform $p(|y|)$ steps on y, the length of $g(y)$ is bounded by $p(|y|)$. This implies that for every y in the range of f there is an x, which is $g(y)$, of length at most $p(|y|)$, such that $f(x) = y$. This is the definition of the honesty of f.   □

The next easy proposition shows that honest functions in PF can be "inverted" in a nondeterministic manner.

Proposition 3.12 *Let f be a polynomially computable, honest function. Then there is a polynomial time nondeterministic machine M which accepts exactly the range of f, and such that, on input y, every accepting computation of M outputs a value x such that $f(x) = y$.*

Proof. Let p be the polynomial provided by the honesty of f. Consider a machine M which performs the algorithm of Figure 3.4.

---

input y
nondeterministically guess x with $|x| \leq p(|y|)$
compute $f(x)$
if $f(x) = y$ then output x and accept
end

Figure 3.4   Inverting an honest function f nondeterministically

---

As f is in PF, the machine M works in polynomial time. An accepting computation of M outputs x only if previously it has been checked that $f(x) = y$. It remains to show that M accepts the range of f. If M accepts, then an inverse image of y has been found; hence y is in the range of f. Conversely, if y is in the range of f, then it follows from the honesty property that some inverse image x of y has length at most $p(|y|)$; thus some computation finds this x and succeeds in proving that $f(x) = y$, therefore accepting y.   □

These propositions allow us now to relate the existence of functions in PF which are not invertible to the $P \stackrel{?}{=} NP$ problem and the honesty properties. First, observe that, by Proposition 3.11, it is enough to consider a dishonest function to show that not every function in PF has its inverse in PF, as indicated in the following example.

Example 3.13 Consider the following function: $f(x) = 1^{\lceil \log_2 |x| \rceil}$. Given any word 1^m in the range of f, the inverse images are those words x such that $\lceil \log |x| \rceil = m$. Therefore, the length of any inverse image of 1^m is exponential in m, and f is not honest. No inverse can be polynomial time computable. However, f is polynomial time computable.

Thus, we restrict ourselves to the honest functions. The question we ask now is the following: are there polynomial time computable honest functions whose inverses are not polynomial time computable? Such functions are called *one-way* functions. The answer is not known. In fact, as we shall see next, this open problem is equivalent to the problem of determining whether $P \stackrel{?}{=} NP$.

Theorem 3.14 $P = NP$ *if and only if every honest partial function computable in polynomial time has an inverse computable in polynomial time.*

Proof. Assume that $P = NP$, and let f be an honest partial function in PF. Let p be the polynomial provided by the honesty of f. Consider the following set (which is in fact a "prefix" set for the inverse of f as in the proof of Theorem 3.9):

$$\text{prefix-inv}(f) := \{ \langle x, y \rangle \mid x \text{ is a prefix of a word } z \text{ such that}$$
$$|z| \le p(|y|) \text{ and } f(z) = y \}$$

The algorithm of Figure 3.5 witnesses the fact that prefix-inv(f) is in NP.

```
input ⟨x, y⟩
guess z with |z| ≤ p(|y|)
check that x is a prefix of z
check that f(z) = y
if both hold then accept
end
```

Figure 3.5   A nondeterministic algorithm for prefix-inv(f)

By hypothesis, $P = NP$, hence prefix-inv(f) is in P. Use the polynomial time deterministic algorithm of Figure 3.6 to invert f.

To prove the converse, let M be a polynomial time nondeterministic machine, and let p be a polynomial such that $p(n)$ symbols are enough to write down a computation of M on any input of length n. We will show that the set accepted by M is in P, and it will follow that $P = NP$. Define the following set:

$$\text{comp}(M) = \{ \langle x, y \rangle \mid x \text{ encodes an accepting computation of } M$$

```
input y
x := λ
loop
  ·  if ⟨x0, y⟩ ∈prefix-inv(f) then x := x0
     else if ⟨x1, y⟩ ∈prefix-inv(f) then x := x1
     else exit
end loop
if f(x) = y then output x
else reject
end
```

Figure 3.6 Inverting an honest function f by "prefix searching"

on input y, and $|x| \leq p(|y|)\}$

Define also the following function:

$$f(\langle x, y \rangle) = \begin{cases} y & \text{if } \langle x, y \rangle \in \text{comp}(M) \\ \textit{undefined} & \text{otherwise} \end{cases}$$

It is immediately clear that $\text{comp}(M)$ belongs to P, and hence that f is computable in polynomial time. Furthermore, if y is in the range of f then there is a pair $\langle x, y \rangle$ such that $f(\langle x, y \rangle) = y$, and the length of $\langle x, y \rangle$ is $O(p(|y|) + |y|)$, which is a polynomial of $|y|$. Therefore f is honest and computable in polynomial time, and by hypothesis it has an inverse g computable in polynomial time.

But, by the definition of inverse, we have that y is in the range of f if and only if $f(g(y)) = y$; this is decidable in polynomial time, because both f and g are computable in polynomial time. Hence the range of f is in P.

Finally note that y is in the range of f if and only if there is an x encoding a computation of M which accepts y, and that this is equivalent to saying that M accepts y. Thus, the set accepted by M is exactly the range of f, which we have shown to be in P. Therefore the set accepted by any nondeterministic polynomial time machine M is in P, from which it follows that $P = NP$.

Hence, one-way functions exist if and only if $P \neq NP$.   □

3.4 Polynomial Time Many-one Reducibility

We introduce the concept of reducibility as a way of comparing the "difficulty" of solving two different problems. This concept allows us to formalize the concept of "the most difficult" elements of a class, and creates a semilattice structure which will be one of the most important sources of intuition about complexity classes. In fact, it will be apparent that a great deal of the results we present throughout this book have a quite direct relationship with this semilattice structure.

Definition 3.15 *Given two sets A_1 and A_2, we say that A_1 is polynomial time many-one reducible to A_2 if and only if there exists a function $f : \Sigma^* \to \Sigma^*$, computable in polynomial time, and such that $x \in A_1$ if and only if $f(x) \in A_2$ holds for all $x \in \Sigma^*$.*

We denote the fact that A_1 is reducible to A_2 by $A_1 \leq_m A_2$, and if f is the function which witnesses the fact that $A_1 \leq_m A_2$ then we will say that $A_1 \leq_m A_2$ via f. Observe that the fact that $A_1 \leq_m A_2$ via f can be expressed equivalently in the following manner: $A_1 = f^{-1}(A_2)$.

We will omit the words "polynomial time" in most of the cases; there should be no confusion, because, unless otherwise stated, all the reducibilities we use in this book are polynomially bounded. We follow the common use of the term "m-reducible" as a shorthand of "many-one reducible".

This reducibility is also called Karp reducibility. Later we shall define other kinds of reducibilities.

In terms of problems, to reduce Problem 1 to Problem 2 is equivalent to finding a function computable deterministically in polynomial time, which, for any instance of Problem 1, constructs an instance of Problem 2 such that it has a solution if and only if the given instance of Problem 1 also has a solution.

We state some interesting properties of the m-reducibility in the next proposition.

Proposition 3.16
(a) $\leq_m$ is a preorder.
(b) $A \leq_m B$ if and only if $\overline{A} \leq_m \overline{B}$.
(c)   For every A, B, $A \leq_m A \oplus B$ and $B \leq_m A \oplus B$.
(d)   For every A, B, C, if $A \leq_m C$ and $B \leq_m C$ then $A \oplus B \leq_m C$.
(e)   If $A \in P$ then, for any other set B different from $\emptyset$ and Σ^*, $A \leq_m B$.

The proof of this proposition is easy, and is left as an exercise (see Exercise 5).

Thus, $\leq_m$ defines an equivalence relation obtained from the preorder in a standard manner: the sets A and B are equivalent (denoted $A \equiv_m B$) if and only if $A \leq_m B$ and $B \leq_m A$. Each equivalence class is called a *polynomial time m-degree*. The polynomial time m-degrees are partially ordered by the order induced by $\leq_m$ on sets. By part (e) of the previous proposition, the class P is the least nontrivial polynomial time m-degree.

Every two polynomial time m-degrees have a least upper bound in the order induced by $\leq_m$: the degree of the join of two sets, one in each m-degree. Indeed, by part (c) of the proposition, the m-degree of the join of two sets is an upper bound on the m-degrees of the sets; further, by part (d), it is the least such upper bound.

Hence the partial ordering of m-degrees is an upper semilattice. It is left to Exercise 7 in this chapter to show that there is no maximum polynomial time m-degree.

However, if we restrict ourselves to considering the degrees contained in a given complexity class, it is possible to find a maximum degree for the class. Such a maximum is identified by following definition, inspired by similar definitions of recursive function theory.

Definition 3.17   *Given a class C,*

1. *A set A is C-hard, or m-hard for C, if and only if, for any set B in C, $B \leq_m A$.*
2. *A set A is C-complete, or m-complete for C, if and only if it is C-hard and $A \in C$.*

The motivation for this definition is that, as we have already stated, we wish to discover the properties which make a given problem difficult. Along this line, it seems worthwhile to define the most difficult problems inside a class, with the hope of isolating the reasons for this difficulty.

Later on in this chapter, after defining other reducibilities, we shall obtain more definitions of completeness. Whenever the reducibility is omitted, the polynomial time m-reducibility should be assumed.

The following proposition establishes some properties of the complexity classes with respect to the m-reducibility and the concept of complete set.

Proposition 3.18   *Let C be any complexity class.*

(a)   *If A is C-hard and $A \leq_m B$ then B is C-hard.*
(b)   *If A is C-complete, $B \in C$, and $A \leq_m B$ then B is C-complete.*
(c)   *If A is C-hard then $\overline{A}$ is co-C-hard.*
(d)   *P, NP, co-NP, and PSPACE are closed under the m-reducibility, i.e. if $A \leq_m B$ and B belongs to some of these classes then A belongs to the same class.*

Again, the proof of this proposition is left as an exercise (see Exercise 6). The concept of a class being "closed" under any other reducibility can be defined in the same straightforward way, and will be used frequently. The *closure* of a class C with respect to a reducibility is the smallest class containing C and closed under the reducibility. Another usual term for referring to the closure of a class C with respect to a reducibility is the *reduction class* of C under the reducibility.

As a corollary to the last proposition, we can deduce the following fundamental properties of complete sets:

Corollary 3.19

(a) *If A is NP-complete or co-NP-complete, and $A \in P$, then $P = NP$.*
(b) *If A is co-NP-complete and $A \in NP$, then $NP = $ co-NP.*
(c) *If A is PSPACE-complete and $A \in P$, then $P = PSPACE$.*

As an example, it follows from the last corollary that it is unlikely that the language formed by all the encodings of prime numbers, which was studied in Theorem 3.4, is NP-complete or co-NP-complete. The reason is that in the same theorem it was shown that both this set and its complement are in NP; if one of them is NP-complete, then the other is co-NP-complete and by part (b) of Corollary 3.19 it would follow that $NP = $ co-NP, which is quite unlikely.

We have defined the concept of NP-complete set, but we have not shown the existence of such sets. "Natural" complete sets will be presented in the next section. We show now that complete sets can be obtained in a rather artificial, but useful, way in the following manner:

Definition 3.20 *Define the following set:*

$$K = \{ \langle M, x, 1^t \rangle \mid M \text{ is a nondeterministic machine}$$

$$\text{that accepts } x \text{ in at most } t \text{ steps } \}$$

In this definition we are identifying the machines M with their encodings over a suitable alphabet.

We show the NP-completeness of K:

Theorem 3.21 *K is NP-complete.*

Proof. First we prove that $K \in NP$, by describing a nondeterministic machine M_1 which accepts K in polynomial time. It is presented in Figure 3.7.

To prove the m-completeness, let A be any set in NP. We have to show that $A \leq_m K$. Let M be a nondeterministic machine which accepts A in polynomial time. Let p be the polynomial that bounds the running time of M, and consider the function $f(x) = \langle M, x, 1^{p(|x|)} \rangle$. The following is true:

- f can be computed in polynomial time. Indeed, M is independent of x, x is copied from the input in linear time, and $p(|x|)$ ones can be written in $p(|x|)$ steps;
- $x \in A$ if and only if x is accepted by M within $p(|x|)$ steps, and this holds if and only if $\langle M, x, 1^{p(|x|)} \rangle \in K$ by the definition of K.

Therefore A is m-reducible to K via f, and the statement of the theorem follows.   □

input $z = \langle M, x, 1^t \rangle$
nondeterministically guess a word w with $|w| \leq t^2$
check that w encodes a sequence of configurations of M, such
    that each of the configurations in the sequence follows
    from the previous one according to M's transition table
check that the initial configuration has x on its input tape
check that the number of configurations is at.most t
check that the last configuration is in an accepting state
if all these conditions hold then accept
end

Figure 3.7   A nondeterministic algorithm for K

3.5 "Natural" *NP*-complete Sets

We must stress the importance of the class *NP*-complete to understand the classes P and *NP*. Observe that by Corollary 3.19, if any polynomial time algorithm is found for some *NP*-complete set, then polynomial time algorithms will be obtained from it for the whole class *NP*, and this class will "fall down" to P. Therefore it seems natural to look for *NP* problems which are *NP*-complete. We will present some examples in this section.

    The first example is the satisfiability problem, SAT, which has been presented in the first section of this chapter. Recall that it consists in determining whether a given boolean formula is satisfiable, and that we have shown in Theorem 3.3 that SAT belongs to *NP*.

    We start our presentation by proving a lemma that shows how to express properties about computations of Turing machines by means of boolean formulas. This lemma will be used again in the next section. We denote by lower case Greek letters such as α, β (with subscripts if necessary) "vectorial" boolean variables consisting of a fixed number of free boolean variables.

Lemma 3.22 *Let M be a single tape Turing machine working in polynomial space. Let p be a polynomial that allows any configuration of M on inputs of length n to be written down with $p(n)$ symbols from the two letter alphabet $\{0, 1\}$. Let α and β be vectorial variables with $p(n)$ free boolean variables each.*

    Then we can construct in polynomial time the following formulas:

(a)   Config(α), *which is true if and only if α receives as value a chain of 0's and 1's which encodes a configuration of M on an input of length n.*

(b)   Next(α,β), *which is true if and only if α and β receive as value chains of 0's and 1's which encode configurations of M of length $p(n)$, and M can go in one step from the first to the second.*

(c)   Equal(α,β), *which is true if and only if α and β receive as value the same chain of 0's and 1's.*

(d)   Initial(α,x), *which is true if and only if α receives as value a chain of 0's and 1's which encodes the initial configuration of M on x;*

(e)   Accepts(α), *which is true if and only if α receives as value a chain of 0's and 1's which encodes an accepting final configuration of M.*

Proof.

(a)   The formula which checks that a string of bits encodes a correct configuration of M is constructed as follows: for each group of bits that may encode a tape symbol, a constant formula is constructed which checks that the symbol is correctly coded. A second constant formula checks whether the symbol contains an indication of the presence of the tape head, and if it does, checks that there is an additional encoding of a correct internal state of the machine. Finally, polynomially many disjunctions check that the input head and the internal state of the machine appear somewhere in the configuration, and polynomially many conjunctions check that the input head does not appear twice in the configuration. The final formula is the conjunction of all these formulas.

(b)   This formula starts from the following conjunction: Config(α)$\wedge$Config(β) A third formula is added which checks that the modification of the symbols and state in the neighborhood of the tape head obeys the transition table of M. Then a fourth formula checks the coincidence of the symbols in the remaining positions of the tape in α and β.

(c)   This formula is constructed by the conjunction of polynomially many "bitwise equality" formulas. This formula is not used in the next theorem, but it is needed in the next section.

(d)   This formula is formed from Config(α), to which an additional formula is added which checks, first, that the input head is at the beginning of the tape; second, that the internal state of M is the initial state; and third, that the contents of the tape is x followed by blanks.

(e)   Again, this formula is formed from Config(α), adding a formula that checks that the internal state is a final accepting state.   □

Note that using a slightly more complicated encoding of configurations by bit strings, similar formulas can be constructed for multitape Turing machines.

Using this lemma we can prove the *NP*-completeness of SAT.

Theorem 3.23   SAT *is NP-complete.*

Proof. It has been shown in Theorem 3.3 that SAT belongs to *NP*. It just remains to show that any other set in *NP* can be m-reduced in polynomial time to SAT.

The proof consists in building a boolean formula from the specifications of a Turing machine and its input, in such a way that there is a satisfying assignment for the formula if and only if there is a polynomially long computation of M which accepts the input.

Let M be a nondeterministic polynomial time machine, and let p be a polynomial which bounds its running time. Let x be an input to M, with $|x| = n$. Combine the formulas constructed in the previous lemma into the formula Accepted(x) as follows:

$$\text{Accepted}(x) = \bigwedge_{i=1}^{p(n)} \text{Config}(\alpha_i) \wedge$$
$$\bigwedge_{i=1}^{p(n)-1} \text{Next}(\alpha_i, \alpha_{i+1}) \wedge \text{Initial}(\alpha_1, x) \wedge \text{Accepts}(\alpha_{p(n)})$$

Such a formula can be written in polynomial time, and is satisfiable if and only if there is a sequence of bit strings to be given as values to the vectorial variables α_i, such that each α_i encodes a configuration, and the machine can follow this sequence of configurations; this means that the sequence is a computation of M. Further, the formula requires the initial configuration to be correct, and the computation to accept. Thus, $x \in L(M)$ if and only if Accepted(x) $\in$ SAT.

Therefore the function mapping each x to the formula Accepted(x) is a polynomial time reduction from $L(M)$ to SAT. Hence SAT is *NP*-complete. □

Once a problem is known to be *NP*-complete, we can add more *NP* problems to the class by showing the existence of a reduction from it. Let us look at the following useful example.

Theorem 3.24 *Let* SAT-CNF *be the set of (encodings of) satisfiable boolean formulas in conjunctive normal form. Then* SAT-CNF *is NP-complete.*

Proof. The same algorithm given for SAT shows that SAT-CNF belongs to *NP*, because it is possible to test in linear time whether a given formula is in conjunctive normal form.

Recall from Theorem 1.35 and Corollary 1.36 that given any boolean formula F, it is possible to construct a formula F' in conjunctive normal form, such that F is satisfiable if and only if F' is satisfiable. An inspection of the proof shows that F' can be constructed from F in polynomial time. Thus, there is a function g in *PF* such that for every formula F, $g(F) = F'$ is a formula in conjunctive normal form, and F is in SAT if and only if $g(F)$ is in SAT-CNF. Thus, SAT-CNF is *NP*-complete by part (b) of Proposition 3.18. □

Today, several hundred problems are known to be *NP*-complete, and there is a column by D. Johnson in the Journal of Algorithms which gives a few new *NP*-complete problems, in every issue. We shall only give one more example of an *NP*-complete problem.

Definition 3.25   *The problem* CLIQUE *is defined as follows: given a graph* $G = (V, E)$, $|V| = n$, *and a positive integer* k, *the problem consists in deciding whether* G *contains a complete subgraph of* k *or more vertices.*

Theorem 3.26   CLIQUE *is NP-complete.*

Proof.   First we must prove that CLIQUE belongs to *NP*. Consider a nondeterministic Turing machine which performs the algorithm of Figure 3.8. This

---

input $G = (V, E)$ and k
guess $V' \subseteq V$
check that $|V'| \geq k$
check that every pair of vertices in V' is joined by an edge of E
if this holds then accept
end

Figure 3.8   A nondeterministic algorithm for CLIQUE

---

nondeterministic computation of M can be done in less than $O(n^2)$ steps.

To prove the *NP*-completeness, we shall reduce SAT-CNF to CLIQUE. As SAT-CNF has been proved *NP*-complete, and CLIQUE is in *NP*, by part (b) of Proposition 3.18 it follows that CLIQUE is *NP*-complete.

Given an instance of SAT-CNF, note that it consists of a set of literals $X = \{x_1, \ldots, x_r\}$, and a set of clauses $\{C_1, \ldots, C_s\}$. The reduction function maps such instance to the following instance of CLIQUE:
$$V = \{\langle x_i, C_j \rangle \mid x_i \text{ occurs in } C_j\}$$
$$E = \{(\langle x_i, C_j \rangle, \langle x_m, C_n \rangle) \mid j \neq n \text{ and } \overline{x}_i \neq x_m\}$$
$$k = s$$
It is left to the reader to prove that the above function is indeed a polynomial time reduction between SAT-CNF and CLIQUE.   □

3.6 "Natural" *PSPACE*-complete Sets

In this section we present an example of a problem which is *PSPACE*-complete. Observe that by Corollary 3.19, all the motivations we have presented for the study of the *NP*-complete sets can also be applied to the *PSPACE*-complete sets.

Our example of a *PSPACE*-complete set is the language QBF presented in Definition 1.34. It corresponds to the following problem: given (an encoding of) a quantified boolean formula without free variables, decide whether it evaluates to "true". The proof that QBF is *PSPACE*-complete will be based on two lemmas. The first one states that QBF is in *PSPACE*. The second one shows how to construct an auxiliary formula which we use later for defining the reduction of any polynomial space machine to QBF.

Lemma 3.27   QBF $\in PSPACE$.

Proof.   In Figure 3.9 we describe a recursive procedure which computes the value of a quantified boolean formula, and the way in which it is called.

---

evaluation-of (F):
    if F is
        the constant *true*: return *true*
        the constant *false*: return *false*
        the negation of F': return the negation of evaluation-of(F')
        a binary boolean operator applied to F' and F'':
            compute evaluation-of(F')
            compute evaluation-of(F'')
            return the result of applying the operator
        $\forall x F'$:
            compute evaluation-of($F'|_{x:=0}$)
            compute evaluation-of($F'|_{x:=1}$)
            return the AND of both results
        $\exists x F'$:
            compute evaluation-of($F'|_{x:=0}$)
            compute evaluation-of($F'|_{x:=1}$)
            return the OR of both results
    end procedure

main program:
input F
$z :=$ evaluation-of (F)
if $z = true$ then accept else reject
end

Figure 3.9   A recursive decision procedure for QBF

If the length of the input is n, then the number of variables plus the number of binary operators is at most n. Therefore the depth of the recursion is at most n. Using a stack for the simulation of the recursive calls, as in the proof of Theorem 2.27, the procedure can be implemented within space n^2. $\qquad\qquad\square$

As in the previous section, we denote by lower case Greek letters "vectorial" boolean variables. Recall from there that, given a machine M working in polynomial space, we can construct in polynomial time the formulas Config(α), Next(α, β), Equal(α, β), Initial(α, x), and Accepts(α), with the meanings discussed there.

Lemma 3.28 *Let M be a Turing machine working in polynomial space, and let m be any nonnegative integer. There exists a quantified boolean formula* Access$\_2^m(\alpha, \beta)$, *with α and β free, which is true if and only if α and β receive as value chains of 0's and 1's which encode configurations of M, and M can go in at most 2^m steps from the first to the second. Furthermore,* Access$\_2^m(\alpha, \beta)$ *can be written by a machine in time polynomial in m.*

Proof.   The proof is by induction on m.

For $m = 0$. The formula

$$\text{Access\_}2^0(\alpha, \beta) = \text{Config}(\alpha) \wedge \text{Config}(\beta) \wedge (\text{Equal}(\alpha, \beta) \vee \text{Next}(\alpha, \beta))$$

fulfills the required conditions.

From $m - 1$ to m. We use the formula Access$\_2^{m-1}(\alpha, \beta)$ provided by the induction hypothesis as follows: M can go from α to β in at most 2^m steps if and only if there is an intermediate configuration γ (exactly in the middle of the computation) such that M can go from α to γ in at most 2^{m-1} steps, and then from γ to β in at most 2^{m-1} steps. Thus, a first attempt leads to the formula

$$\exists \gamma (\text{Config}(\gamma) \wedge \text{Access\_}2^{m-1}(\alpha, \gamma) \wedge \text{Access\_}2^{m-1}(\gamma, \beta))$$

However, it is easy to see that this formula is too long for our purposes, because it has a length of $\Omega_\infty(2^m)$.

Nevertheless, we can use the fact that both sides of the last conjunction have the same format, and state it only once. We use a universal quantifier to ensure that both the accessibility of γ from α and of β from γ are asserted. The required formula is:

$$
\begin{aligned}
\text{Access\_}2^m = \exists \gamma \text{Config}(\gamma) \wedge \\
\forall \alpha', \forall \beta'[(\text{Equal}(\alpha', \alpha) \wedge \text{Equal}(\beta', \gamma)) \vee \\
(\text{Equal}(\alpha', \gamma) \wedge \text{Equal}(\beta', \beta)) \Rightarrow \\
\text{Access\_}2^{m-1}(\alpha', \beta')]
\end{aligned}
$$

The length of this formula is linear in m, and its structure is simple enough to be written down in $p(m)$ steps for some (small) polynomial p.   □

Now we can prove the completeness result mentioned above.

Theorem 3.29   QBF *is PSPACE-complete.*

Proof. By Lemma 3.27, QBF is in *PSPACE.* Thus, it is enough to show that it is *PSPACE*-hard. Once more, the proof consists in building a quantified boolean formula from the specifications of a Turing machine and its input, in such a way that the formula is true if and only if M accepts the input in polynomial space.

Let M be a polynomial space Turing machine, and let p be its space bound. Notice that if a word x of length n is accepted by M then, by the argument given in Lemma 2.25, the accepting computation has at most $2^{c \cdot p(n)}$ configurations.

For each input x to M, with $|x| = n$, consider the formula Accepted(x) constructed as follows:

$$\text{Accepted}(x) = \exists \alpha \exists \beta (\text{Initial}(\alpha, x) \wedge \text{Accepts}(\beta) \wedge \text{Access\_2}^{c \cdot p(n)}(\alpha, \beta))$$

By the previous lemma, this formula can be constructed in time polynomial in n, and evaluates to "true" if and only if the machine M accepts x, so that $x \in L(M)$ if and only if Accepted(x) $\in$ QBF.

Thus, the function mapping each x to the formula Accepted(x) is a reduction from $L(M)$ to QBF. Hence QBF is *PSPACE*-complete. $\qquad\square$

3.7 Padding Arguments

In this section we show several inequalities among complexity classes, and also several implications among inclusions of complexity classes. All have in common that they are shown by a similar argument: "padding" a hard language by appending to its elements an easy-to-recognize "tail". This process increases artificially the length of the inputs, and thus the "padded language" is easier to recognize than the initial one.

Two continuations are possible for the argument. In one of them, a closure under polynomial time m-reducibility can be exploited (although not always explicitly) to show an implication between two inclusions of complexity classes. This is the case for Theorem 3.30 below. In the other one, via this "padding" process, inclusions among "easy" classes translate into inclusions among "harder" ones, as in Theorem 3.32.

We will present only two major theorems along this line, together with some corollaries. Other results are proposed in the exercises, and many others have been reported in the references listed in the Bibliographical Remarks.

Theorem 3.30   *If DSPACE(n) $\subseteq$ P, then P = PSPACE.*

Proof. Assume that DSPACE(n) $\subseteq$ P, and let L be a set in *PSPACE.* Let M be a machine accepting L in space $p(n)$ for some polynomial p. Consider the following set:

$$L' = \{w10^{p(|w|)} \mid w \in L\}$$

Then a machine M' which, first, scans its input looking for the rightmost 1, and then simulates M on the word w preceding it, accepts L' in linear space. Therefore, by hypothesis, $L' \in P$. Let M'' be a machine accepting L' in polynomial time. Then another machine which on input w appends to it a "tail" $10^{p(|w|)}$ and then simulates M'' on the word obtained accepts L in polynomial time.   □

In fact, there is in the preceding proof a "hidden" use of the fact that P is closed under m-reducibility: the key to the proof is that for every set L in $PSPACE$ there is another set L' in $DSPACE(n)$ such that L is m-reducible to L'. Thus, if $L' \in P$ then by the closure of P under polynomial time m-reducibility we get $L \in P$. As a corollary of the preceding theorem we obtain the following interesting fact.

Corollary 3.31 $P \neq DSPACE(n)$.

Proof.   Assuming $P = DSPACE(n)$ implies $P = PSPACE$ by the previous theorem. Therefore $PSPACE = DSPACE(n)$, which contradicts the Space Hierarchy Theorem (Theorem 2.29).   □

It should be noticed that the preceding theorem and corollary are easily seen to be true for any other class of the form $DSPACE(n^r)$ for every real $r > 0$. The whole argument remains valid. Exercises 10 and 11 require a similar proof.

We illustrate next the use of padding techniques for deriving inequalities in lower classes from inequalities in upper classes. These kinds of results are known as "downward separations", and they have sometimes been expressed under the form of "translational lemmas". The statement of our next theorem is exactly the contrapositive of a downward separation; this statement makes the proof more transparent.

Theorem 3.32   *If $P = NP$, then for every time constructible function f, both $DTIME(2^{O(f)}) = NTIME(2^{O(f)})$ and $DTIME(f^{O(1)}) = NTIME(f^{O(1)})$ hold.*

Proof.   We show that if $P = NP$ then $DTIME(2^{O(f)}) = NTIME(2^{O(f)})$. The other statement is proved analogously. The inclusion from left to right is known from Theorem 2.26. To show the converse inclusion, let L be any set in $NTIME(2^{O(f)})$, and let M be a nondeterministic machine accepting it in time $2^{c \cdot f(n)}$ for some fixed constant c. Define a padded version of L as in Theorem 3.30:

$$L' = \{w 10^{2^{c \cdot f(|w|)}} \mid w \in L\}$$

Then a nondeterministic machine can accept L' in linear time, searching for the rightmost 1 and simulating M on the word w at the left of this 1. By hypothesis, L' is in P. Let M' be a deterministic polynomial time machine accepting L'. From it, we obtain a deterministic machine accepting L in time $2^{c' \cdot f(n)}$, which first appends to its input w a suitable "tail" $10^{2^{c} f(|w|)}$ and then simulates the polynomial time deterministic machine on the padded input. Thus, L is in $DTIME(2^{O(f)})$, as was to be shown.   □

Corollary 3.33   *If $P = NP$ then $DEXT = NEXT$.*

In the next chapter we will present a different proof of this last corollary. Similar translations hold among many complexity classes. Exercises 12 and 13 present some of them. See the Bibliographic Remarks for more information on the subject.

3.8 Space Bounded Reducibility

The polynomial time m-reducibility introduced in the previous sections is useful for obtaining complete problems for classes defined by resource bounds greater than polynomials, like NP, $PSPACE$, or $DEXT$. But it is not appropriate to deal with classes like P or $NLOG$, since the resources which define these classes are "finer" than deterministic polynomial time.

Thus, we present next the space bounded m-reducibility, in a general setting. The particular case of the logarithmic space reducibility will be appropriate for defining completeness of these complexity classes. In particular, this reducibility will play an important role in Volume II, when dealing with parallel complexity classes.

Unless otherwise stated, all the space bounding functions in this section are to be assumed greater than or equal to $\log n$.

Definition 3.34   *Given two sets $A_1 \subseteq \Sigma^*$ and $A_2 \subseteq \Sigma^*$, we say that A_1 is s-space reducible to A_2 if there is a function $f : \Sigma^* \to \Sigma^*$, such that:*

1. *$f(x)$ is computable in space $s(|x|)$;*
2. *for all x in Σ^*, $x \in A_1$ if and only if $f(x) \in A_2$;*
3. *there is a positive integer c such that for all x in Σ^*, $s(|f(x)|) \leq cs(|x|)$.*

We denote such a reduction by $\leq_m^s$. As we already mentioned, an important particular case is when s is the log function, which will be called the *log-space reducibility* and denoted $\leq_m^{\log}$.

Condition 1 in the definition of space bounded reducibility is natural, and Condition 2 states just the fact that f computes a reduction. Condition 3 is less intuitive, and has to do with the major problem of the space bounded reducibility: the transitivity. Naturally, we desire our reducibilities to be transitive; however, the fact that a space bounded machine may write an output substantially larger than allowed by its space bound may hide several difficulties when we try to compute the composition of two functions. Let us briefly discuss these difficulties.

First, assume that f_1 and f_2 are functions computed in space $s(\ )$ on their input, and assume that we wish to compute the composition $f_2 \circ f_1$. Assume that on input x of length n, f_1 yields as output a long result, say of length $h(n)$. Now, the fact that f_2 can be computed in space $s(\ )$ means that to find $f_2 \circ f_1(x)$ we could need as much as $s(h(n))$ space, which is required by f_2. For certain functions s, the value $s(h(n))$ might be larger than $s(n)$, and this makes it absolutely impossible to compute the composition of both functions in space $s(n)$. Condition 3 is imposed precisely to overcome this problem, since it says that the space bound for the length of x is linearly related to the space bound for the length of $f(x)$. It should be mentioned that for the most interesting case, $s(n) = \log n$, Condition 3 always holds; the reader can check it in a straightforward way.

But there is still another subtlety involved in the composition of functions computable in bounded space. Let f_1 and f_2 be as before, and consider again their composition. To compute this composition in space s, the naive simulation consisting in computing f_1, storing the result in a work tape, and using it as input for computing f_2, does not work. Indeed, the length of $f_1(x)$ may be larger than $s(|x|)$, and therefore keeping the intermediate result in a work tape may overrun the allowed space bound. Instead, we use a different technique to compute in bounded space the composition of functions. Intuitively, this can be described as "compute it again every time you need it". In fact, the idea is to simulate the whole computation of the first machine every time the second machine requires a new input symbol. It is described more formally in the next result.

Lemma 3.35 *Let M_1 be a transducer computing the function f_1 in space s, and let M_2 be another transducer computing f_2 within space t. Then there exists a transducer M_3 computing the composition $f_2 \circ f_1$ within the space bound*

$$\max\{k \cdot s(|x|), t(|f_1(x)|)\}$$

where k is a constant depending only on M_1.

Proof. Construct M_3 as follows. For any given input x, M_3 will simulate the effect of M_2 applied to the input $f_1(x)$. As indicated above, this cannot be done directly, because $f_1(x)$ may be too large to be stored within space $s(|x|)$.

Instead, the output of M_1 is fed directly to M_2 in the following way: M_3 uses a subroutine which, when given any integer i ($0 \leq i \leq |f_1(x)| + 1$), will return the ith symbol of $f_1(x)$, thus simulating the effect of reading the ith symbol of the input tape of M_2. To do this, the computation of M_1 is simulated from the very beginning every time M_2 requires to read a symbol from the input tape. The algorithm is presented in Figure 3.10.

The machine M_3 executing this algorithm requires $O(s(|x|))$ space to perform (repeatedly, but reusing the same space) the simulation of M_1, and

produce-symbol(i):
    resume the simulation of M_1 on x
        starting at the initial configuration
    during the simulation, count the number of symbols
        written by M_1 on the output tape
    when the count reaches i, stop the simulation
    return the ith symbol written by M_1

main program:
input x
$i := 1$
loop
    produce-symbol(i)
    using the symbol obtained by the subroutine,
        simulate one step of M_2
    if in this step M_2 moves its input head left
        then $i := i - 1$
    else if in this step M_2 moves its input head right
        then $i := i + 1$
until M_2 stops
accept or reject according to M_2
end

Figure 3.10 Computing the composition of two functions

$O(t(|f_1(x)|))$ to simulate M_2. Besides, some counters are needed to record the position of the simulated input head of M_2 and the position of the output tape head of M_1. Keeping these counters in binary, it is easy to check that the indicated amount of space is enough. $\qquad\square$

This lemma can be applied as well to the case when M_2 is an acceptor instead of a transducer.

Corollary 3.36 *Let M_1 be a transducer computing a function f in space $s(|x|)$, and let M_2 be a deterministic (nondeterministic) machine working with space bound $t(|y|)$. Then there exists a deterministic (nondeterministic) machine M_3 working in space*

$$\max\{k\cdot s(|x|), t(|f(x)|)\}$$

which accepts x if and only if M_2 accepts $f(x)$.

We can prove next two important properties of the space-bounded reducibility.

Theorem 3.37 *The relation $\leq_m^s$ is reflexive and transitive.*

Proof. The reflexivity follows from the fact that the identity function is computable using constant work space. To prove the transitivity, let A, B, and C be sets such that $A \leq_m^s B$ and $B \leq_m^s C$. By definition, there exist transducers M_1 and M_2, constants c_1 and c_2, and functions f_1 and f_2 which are computable by M_1 and M_2 within space s, and such that for all x in Σ^*,

$$s(|f_2(f_1(x))|) \leq c_1 \cdot s(|f_1(x)|) \cdot \leq c_2 \cdot c_1 \cdot s(|x|)$$

Therefore, $f_2 \circ f_1$ fulfills Conditions 2 and 3 in the definition of space reducibility, having $c_2 \cdot c_1$ as a suitable constant for the bound on the output length. To prove that $f_2 \circ f_1$ is indeed computable in space s, let M_3 be the transducer constructed in Lemma 3.35. It is easy to see that M_3 computes $f_2 \circ f_1$ within space $s(|x|)$. $\square$

The transitivity of the space bounded reducibility implies that we can define in the canonical way the equivalence relation $\equiv_s$, providing a definition of the concept of complete language as we have done in previous sections.

Several natural complexity classes are closed under space bounded reducibility.

Theorem 3.38 *Let $A \leq_m^s B$. Then*

(a) *If B belongs to DSPACE(s) then A belongs to DSPACE(s).*
(b) *If B belongs to NSPACE(s) then A belongs to NSPACE(s).*

Proof. We prove part (a); the proof of part (b) is identical. Let f be the reduction from A to B. Then f is computed by a transducer M using space s, and $s(|f(x)|) \leq c \cdot s(|x|)$ for any $x \in \Sigma^*$ and for some constant c. Let M_2 be a deterministic machine accepting every y in B in space $(s|x|)$. The machine M_3 obtained from M_1 and M_2 as indicated in Corollary 3.36 recognizes A within space $O(s(|x|))$. $\square$

As we anticipated in the introduction to this section, an especially important case is the log-space reducibility $\leq_m^{\log}$. Using this reducibility, we obtain part (b) of Corollary 3.39 as a particular case of the previous result. In a similar manner, the closure of time bounded classes under space bounded reducibility can also be shown. The only difference is that the argument presented in Lemma 2.25 about time bounds for space bounded machines must be taken into account. Part (a) of Corollary 3.39 is a result of this form.

Corollary 3.39
(a) *If $A \leq_m^{\log} B$ and $B \in P$ then $A \in P$.*
(b) *If $A \leq_m^{\log} B$ and $B \in NLOG$ then $A \in NLOG$.*

The log-space reducibility will be very useful in Volume II. It allows us also to define complete sets for $NLOG$ and P. A problem that is log-space complete for $NLOG$ is presented in Exercise 16, and a problem which is log-space complete for P will be presented in the exercises of Chapter 5. For other complete problems for $NLOG$ and P, the interested reader may consult the references indicated at the end of the chapter.

3.9 Exercises

1. Let f be a function in PF. Show that for every x, the set $f^{-1}(x)$ is in P. Generalize your answer to all of the sets $f^{-1}(A)$ where A is in P.

2. Show that a set is in NP if and only if it is the range of an honest polynomially computable function. Show that a set is in P if and only if it is the range of an invertible polynomially computable function.

3. Consider the following alternative definition of an honest function, which can sometimes be found in the literature: the function f is honest if and only if $f(x)$ can be computed in time $p(|f(x)|)$, for some polynomial p. Explain (and prove) the relationship between this definition and the definition given in the text,

   (a) for all partial functions from Σ^* to Σ^*;
   (b) for the polynomial time computable, one-one partial functions from Σ^* to Σ^*.

4. A *single-valued nondeterministic Turing machine* is a nondeterministic Turing machine with output tape, such that on each accepted input x, the output of every accepting computation is the same. The class of partial functions $NPSV$ consists of all the partial functions computed by single-valued nondeterministic Turing machines working in polynomial time. Show that $P = NP$ if and only if $PF = NPSV$.

5. Prove Proposition 3.16.

6. Prove Proposition 3.18.

7. Prove that in the semilattice of the m-degrees there is no maximum polynomial m-degree.

8. Show that if C is a class which is closed under complements, and A is a C-hard set with respect to the m-reducibility, then $\overline{A}$ is also C-hard.

9. The *polynomial time nondeterministic m-reducibility* is defined as follows: $A \leq_m^{NP} B$ if and only if there is a polynomial time nondeterministic machine M with output tape such that for every x, $x \in A$ if and only if some computation of M on input x produces an output $y \in B$. Which of the statements of Proposition 3.16 hold for this nondeterministic polynomial time m-reducibility? Prove your answer.

10. Show that if $NSPACE(n) \subseteq NP$ then $NP = PSPACE$. Conclude that $NSPACE(n) \neq NP$.

11. Show that if $DTIME(2^{c \cdot n}) \subseteq NP$ for some c then

$$DEXT \subset NP = PSPACE = EXPTIME$$

and that the first inclusion is strict. Conclude that for every c

$$DTIME(2^{c \cdot n}) \neq NP$$

12. Show that if $P = PSPACE$ then, for every time constructible function f, $DTIME(2^{O(f)}) = DSPACE(2^{O(f)})$, and $DTIME(f^{O(1)}) = DSPACE(f^{O(1)})$.

13. Show that if $LOG = NLOG$ then, for every space constructible function f with $f(n) \geq \log(n)$, $DSPACE(f) = NSPACE(f)$.

14. *Show that $DEXT \neq PSPACE$. *Hint:* Show that DEXT is not closed under m-reducibility.

15. *Consider the set CSG-Acceptance, defined as follows: it consists of the pairs $\langle g, x \rangle$ where g encodes a context-sensitive grammar that generates x. Show that this set is PSPACE-complete. If necessary, check in any textbook on theoretical computer science the definition of context-sensitive grammar and the equivalence of this model with linear bounded automata.

16. *Given a directed graph $G = (V, E)$, $V = \{1, 2, \ldots, n\}$, the *graph reachability problem* is to determine if there is a path in G from vertex 1 to vertex n. Prove that the graph reachability problem is $NLOG$-complete. *Hint:* To prove that it is in $NLOG$, notice that it is not necessary to store the whole path. To prove completeness, fix a nondeterministic logarithmic space machine M and its input x and construct the graph G_x of all the configurations of M on x.

3.10 Bibliographical Remarks

The development of the complexity classes according to the use of resources, particularly time and space, soon leads to consideration of the classes presented in this chapter. In particular, the first works to devote some attention to the class P (although this name was coined later) are Cobham (1964) and Edmonds (1965). Eventually the names P and NP were coined by Karp (1972). The proof that PRIMES is in NP (Theorem 3.4) is due to Pratt (1975). A deterministic algorithm that tests for primality in polynomial time has been devised by Miller (1976). However, the proof of its correctness relies on the still open Extended Riemann Hypothesis, and therefore does not constitute a proof that PRIMES $\in P$. For a survey of subexponential algorithms for primality, see the book by Kranakis (1986).

There is no full agreement on the notation for exponential time classes. The name $EXPTIME$ has been used sometimes in the literature for the class we denote $DEXT$, and other authors use it with the same meaning as we do here. Since $DEXT$ has been used only with one meaning, we have chosen EX-$PTIME$ to denote the other class. Note that they are mnemonics for, respectively, **Deterministic EX**ponential (linear exponent) **T**ime and **EX**ponential (Polynomial exponent) TIME, and that no "P" appears in $DEXT$.

Many parts of the material presented here about invertibility and honesty of functions have been inspired by the Ph. D. Dissertation of E. Allender (1985), where some nice considerations on the applications of invertible functions can be found. See also Valiant (1976). Other complexity classes of functions based on nondeterministic Turing machines are defined in Valiant (1979).

The semilattice structure of the m-degrees was studied by Ladner (1975a), where he proved the existence of recursive sets A and B not in P such that $A \leq_m B$ but $B \not\leq_m A$. From the technique used follow several important consequences, such as for example the density of the partial order defined by the reducibilities. We shall present such results in Chapter 7 by use of other techniques. We shall mention more about this structure in the next chapter, where we study the T-reducibility.

In the same way as was done in recursion theory, Cook (1971) and Karp (1972), defined the concept of "most difficult" problem in the class NP, the NP-complete problems, by use of the concept of reducibility. Cook used the Turing reducibility (see Chapter 8 of Rogers (1967)), while Karp used the many to one reducibility (see Chapter 7 of Rogers (1967)). The proof of the NP-completeness of SAT is due to Cook (1971), while the formal definition of the class NP-complete is from Karp (1972). An excellent survey of the NP-completeness results is the book by Garey and Johnson (1978). Cook's reducibility will be studied in the next chapter. This research was conducted mainly in the United States. It must be mentioned that, independently and almost simultaneously, the concept of "brute force search" (or *perebor*) led the Russian mathematicians to the same concepts. In particular, Levin defined the NP-completeness in this context. The reader may wish to read the very interesting survey by Trakhtenbrot (1984).

The first known $PSPACE$-complete problems appear in Karp (1972). The result presented here about the $PSPACE$-completeness of QBF is due to Stockmeyer and Meyer (1973).

Padding techniques have been known for quite a long time. Among the many authors deriving results in this way, let us mention Book (1972), where Theorem 3.30, Corollary 3.31, and Exercises 10 and 11 appear; Hartmanis and Hunt (1974), where a weaker version of Theorem 3.32 appears; Book (1974a), where we have found a full version of Theorem 3.32 and also Exercises 12 and 14; and finally Savitch (1970) and Simon (1975), where Exercise 13 can be found. These references provide many other results along the same line. A more general framework in which padding techniques are presented, together with some new results, is Book (1976).

The space bounded reducibility was studied in detail by Jones (1975). Our presentation follows this work. Problems log-space complete for P are presented by, among others, Cook (1973), Jones and Laaser (1976), and the above-mentioned Jones (1975). Log-space complete problems for $NLOG$ are considered in Savitch (1970), from where we have taken Exercise 16; Jones, Lien, and Laaser (1976), and others. The book by Hopcroft and Ullman (1979) presents all its complexity theory using the log-space reducibility.

4 Time Bounded Turing Reducibilities

4.1 Introduction

After introducing the polynomial time m-reducibility and the logarithmic space m-reducibility in Chapter 3, in this chapter we shall introduce other types of reducibilities and related considerations, like the properties of sets of low density and the concept of self-reducible set.

The first section is on the polynomial time T-reducibility and the relativizations of complexity classes. The next section gives a glimpse of the "density" question and its influence on the relationships between several complexity classes. The last two sections study, respectively, a more general form of reducibility known as SN-reducibility, and the so-called "self-reducible sets". The material in all these sections is crucial for several sections of other chapters as well as for part of Volume II.

Some other topics of great interest are treated in the exercises; for instance, other forms of reducibility are defined there. Quite a few of the results indicated in these exercises are used later in the exercises of subsequent chapters.

4.2 Polynomial Time Turing Reducibility: Relativized Classes

We define now other types of reducibility. In particular, we present the polynomial time version of Turing reducibility, and show how it leads to a very important concept: the relativized complexity classes. This reducibility, as well as the concept of relativized class, is based upon the model of the oracle machine which was defined in Chapter 1.

Definition 4.1 *Given two sets A_1 and A_2, we say that A_1 is polynomial time Turing reducible to A_2 if and only if there exists a deterministic polynomial time oracle machine M such that $A_1 = L(M, A_2)$.*

We denote the fact that A_1 is Turing reducible to A_2 in polynomial time by $A_1 \leq_T A_2$. If $A_1 \leq_T A_2$, and the machine M witnesses this fact, we will say that $A_1 \leq_T A_2$ *via* M.

Again, as we do with the m-reducibility, we will omit the words "poly-nomial time" in most of the cases. We use the term "T-reducible" as a shorthand for "Turing reducible". This polynomial time Turing reducibility has also been called Cook reducibility.

In terms of problems, to reduce Problem 1 to Problem 2 by a Turing re-ducibility is equivalent to finding a deterministic algorithm in which calls are made to a "subroutine" which solves instances of Problem 2, and demanding the algorithm work in polynomial time; the "subroutine calls" are supposed to take unit time.

Thus, even if Problem 1 is not solvable in polynomial time, because solv-ing it presents some kind of difficulty, it is possible to find some instances of Problem 2 that give us some additional information. This extra information allows us to overcome the difficulties and decide Problem 1 in polynomial time. In some sense, Problem 2 "contains enough information" to solve also Problem 1.

Before going on to the study of other properties of this reducibility, we show an example of such a reducibility.

Example 4.2   Let MAXCLIQUE be the following problem: given a graph $G = (V, E)$, $|V| = n$, and a positive integer k, the problem consists in deciding whether the maximal complete subgraphs of G are of size k. Such a set is reducible to CLIQUE in the following way. Consider a deterministic oracle machine M which works as described in Figure 4.1.

---

input G, k
query the oracle CLIQUE about (G, k)
if the answer is YES then
    query the oracle CLIQUE about $(G, k + 1)$
    if the answer is NO then accept
in any other case reject

Figure 4.1   A Turing reduction from MAXCLIQUE to CLIQUE

It is immediately seen that this is a polynomial time Turing reduction from MAXCLIQUE to CLIQUE.

In the next proposition, we state some properties of the T-reducibility.

Proposition 4.3
(a) $\leq_T$ is a preorder.
(b) $A \leq_T B$ if and only if $\overline{A} \leq_T B$, if and only if $A \leq_T \overline{B}$, if and only if $\overline{A} \leq_T \overline{B}$.
(c)   For every A, B, $A \leq_m B$ implies $A \leq_T B$. Hence, for every A, B, $A \leq_T A \oplus B$ and $B \leq_T A \oplus B$.

(d) *For every A, B, C, if $A \leq_T C$ and $B \leq_T C$ then $A \oplus B \leq_T C$.*

(e) *If $A \in P$ then, for any other set B, $A \leq_T B$.*

The proof of this proposition is left as an exercise (see Exercise 1), and is quite similar to the solution of Exercise 5 of Chapter 3.

Thus, in the same manner as for $\leq_m$, we found that $\leq_T$ defines an equivalence relation obtained from the preorder. We denote the fact that the sets A and B are equivalent by $A \equiv_T B$. Each equivalence class is called a *polynomial time T-degree*. Again, the polynomial time T-degrees are partially ordered by the order induced by $\leq_T$ on sets, and the class P is the least polynomial T-degree. This partial ordering of T-degrees is also an upper semilattice, in which the least upper bound is given by the degree of the join.

As we did for the m-reducibility, it is interesting to formalize the concept of "most difficult" set in a class. Indeed, some problems can be shown to be "universal" for a class; this means that they contain "enough information" to solve *any* problem in the class. This notion is formalized by requiring that every problem in the class be reducible to the "universal" problem. For example, this is the case for the problems shown to be m-complete for NP and for $PSPACE$ in the previous sections.

Thus, we can consider hard and complete sets for a class in terms of Turing reducibility.

Definition 4.4 *Given a class C,*

1. *A set A is C-T-hard, or T-hard for C, if and only if for any set B in C, $B \leq_T A$.*

2. *A set A is C-T-complete, or T-complete for C, if and only if it is C-T-hard and $A \in C$.*

The following proposition is proved in exactly the same manner as Proposition 3.18 (Exercise 6 of Chapter 3).

Proposition 4.5 *Let C be any complexity class.*

(a) *If A is C-T-hard and $A \leq_T B$ then B is C-T-hard.*

(b) *If A is C-T-complete, $B \in C$, and $A \leq_T B$, then B is C-T-complete.*

(c) *A is C-T-hard if and only if $\overline{A}$ is C-T-hard.*

(d) *P and $PSPACE$ are closed under the T-reducibility.*

Note that it is not known whether NP is closed under the T-reducibility. See Exercise 4. Nevertheless, we can obtain a corollary similar to Corollary 3.19 from the proposition.

Corollary 4.6
(a) *If A is NP-T-complete or co-NP-T-complete, and $A \in P$, then $P = NP$.*
(b) *If A is PSPACE-T-complete and $A \in P$, then $P = PSPACE$.*

Thus, to each set A we can associate a class of sets: those sets that can be solved in polynomial time given A as oracle. Note that when A is the empty set, or any other set in P, we have that this class is P. Thus, we introduce the following notation.

Definition 4.7 *Given any set A, the class $P(A)$ consists of all the sets that are T-reducible in polynomial time to A.*

This class is called a "relativization of P to the oracle A". Of course, there is no reason to stop here. We can use nondeterministic machines to define the following:

Definition 4.8 *A is polynomial time nondeterministic Turing reducible to B (denoted $A \leq^{NP} B$) if and only if $A = L(M, B)$ for some nondeterministic polynomial time oracle machine M.*

Some properties of this reducibility are proposed in Exercises 8 and 9. As we did with the deterministic reducibility, it is possible for us to consider the class of sets that are nondeterministically T-reducible in polynomial time to A.

Definition 4.9 *Given any set A, the class $NP(A)$ consists of all the sets that are nondeterministically T-reducible in polynomial time to A.*

In this way, any complexity class can be relativized to any oracle set. We can obtain "replications" of all our complexity classes starting with any set A, and these classes are said to be "relativized to A".

Definition 4.10
1. *$LOG(A)$ is the class of sets which can be decided in deterministic logarithmic space with an oracle for A.*
2. *$NLOG(A)$ is the class of sets which can be decided in nondeterministic logarithmic space with an oracle for A.*
3. *$PSPACE(A)$ is the class of sets which can be decided in deterministic polynomial space with an oracle for A.*
4. *$DEXT(A)$ is the class of sets which can be decided in deterministic exponential time with an oracle for A.*

Similarly, we get $NEXT(A)$, $EXPSPACE(A)$, and so on. In later parts of this book we will study some of these classes in depth. For the moment, we just want to point out that the same kind of concept we have been dealing with up to now can be translated to the relativized classes. We will deal specifically with classes $P(A)$, $NP(A)$, and $PSPACE(A)$. The following properties of these classes are easy to prove (see Exercise 7), and many other properties of the unrelativized classes carry over to every relativization.

Proposition 4.11   *For any set A,*

(a) $P(A)$ *is included in* $NP(A)$.
(b) $A \in P(A)$, *and hence* $A \in NP(A)$.
(c) $P(A) = P(\overline{A})$ *and* $NP(A) = NP(\overline{A})$.
(d) $P(A)$ *is closed under complementation.*
(e) $P(A) = P(B)$ *if and only if* $A \equiv_T B$.
(f) $P(A)$ *and* $NP(A)$ *are closed under* $\leq_m$.
(g) $NP(A)$ *is included in* $PSPACE(A)$.
(h) *If* $A \in P$ *then* $P(A) = P$.

The above notation can be generalized in the following way:

Definition 4.12   *Given any class of sets C, the class $P(C)$ consists of all the sets that are T-reducible in polynomial time to some set in C; and the class $NP(C)$ consists of all the sets that are nondeterministically T-reducible in polynomial time to some set in C.*

Notice that, in this way, the classes P and NP can be considered as operators among classes, in the sense that given a class C we can apply to it the operator $P(\ )$ obtaining the class $P(C)$, and the same can be done with NP. For instance, Part (h) of Proposition 4.11 can be rephrased by saying that $P(P) = P$. An interesting relationship between them is given in the following result.

Proposition 4.13   *For any class C, $NP(P(C)) = NP(C)$.*

Proof.   Since C is included in $P(C)$, the inclusion from right to left is immediate. To prove the converse inclusion, let B be a set in $NP(P(C))$. By definition, there exists a set D in $P(C)$ such that B is in $NP(D)$. Therefore, there is a deterministic Turing machine M_1 and a set A in C such that $D = L(M_1, A)$, and there exists a nondeterministic Turing machine M_2 such that $B = L(M_2, D)$.

Let M_3 be a machine which works as indicated in Figure 4.2.

It is easy to see that the running time of M_3 is polynomial, and that $L(M_3, A) = B$. Hence B is in $NP(C)$.   □

It is left to the reader (Exercise 17) to state and prove an analogue to Proposition 4.11 for classes of the form $P(C)$ or $NP(C)$, where C is a complexity class. We will state just the following relationship between the relativization of a class and those of their complete sets.

input x
loop
    nondeterministically, simulate the computation of
        M_2 on x until it stops or queries the oracle
    if M_2 stops
        then exit from the loop
        else let w be the queried word
    simulate the computation of M_1 on w with oracle A
    if M_1 accepts w
        then continue the simulation of M_2 in the YES state
        else continue the simulation of M_2 in the NO state
end loop
if M_2 accepts then accept else reject
end

Figure 4.2   Algorithm for the proof of Proposition 4.13

Proposition 4.14   *Let C be any class of sets. If A is C-T-complete then $NP(C) = NP(A)$ and $P(C) = P(A)$.*

Proof.   Since $A \in C$, by definition $NP(A) \subseteq NP(C)$ and $P(A) \subseteq P(C)$.

To prove the converse inclusion, let $B \in NP(C)$. This implies that there exists a set $D \in C$ such that $B \in NP(D)$. But as A is C-T-complete, we know that $D \leq_T A$, i.e. $D \in P(A)$. Therefore $B \in NP(P(A))$, which is $NP(A)$ by Proposition 4.13.

The case of P is analogous, using the transitivity of $\leq_T$.   □

Again we can define complete problems for the relativized classes. We continue to identify Turing machines with their encodings over some suitable alphabet. Define the following set:

Definition 4.15   *For any set A,*

$$K(A) = \{\langle M, x, 1^t \rangle \mid M \text{ is a nondeterministic machine}$$

$$\text{that accepts } x \text{ with oracle } A \text{ in at most } t \text{ steps} \}$$

In an analogous manner to Theorem 3.21, using nondeterministic oracle machines for M and M', it can be shown that the following holds:

Theorem 4.16   *$K(A)$ is m-complete for $NP(A)$.*

The proof is left to the reader. See Exercise 19. A similar version using deterministic machines can be shown to be m-complete for $P(A)$, as proposed in the same Exercise 19.

4.3 Tally and Sparse Sets in *NP*

Let us use the concepts of sparse and tally sets to study some relationships between the classes we have defined. Recall that the definitions of tally and sparse were given in Chapter 1. Recall also that we are identifying words of $\{0,1\}^*$ with natural numbers, via a binary coding. We will use the same notation for a number and for the word coding it.

Part of the interest of tally sets comes from the following transformation, which constructs a tally set containing the same information as a given set, but in a "spread" form, so that the information is kept "exponentially away":

Definition 4.17 *Given a set A in $\{0,1\}^*$, we define its tally version as follows:*

$$\text{tally}(A) = \{0^n \mid n \in A\}$$

where n is taken as a number in the expression 0^n and as the corresponding binary word in the expression $n \in A$ (see Section 1.2).

Observe that the length of 0^n is n, which is exponential in the length of the binary codification of n. Therefore, we can establish the following relationships between the complexity of a set and the complexity of its tally version.

Proposition 4.18
(a) $A \in DEXT$ *if and only if* $\text{tally}(A) \in P$.
(b) $A \in NEXT$ *if and only if* $\text{tally}(A) \in NP$.

Proof.   Assume that A is in *DEXT*. Let M be a deterministic exponential time bounded machine which accepts A. Consider the machine M' of Figure 4.3.

---

input w
if w is not of the form 0^n then reject
else

    simulate M on input n
    accept if and only if M accepts
end

Figure 4.3   Deciding the tally version of a set

It is immediately clear that M' accepts the tally version of A. Its running time is exponential in $|n|$, i.e. $2^{c \cdot |n|}$. But $|n|$ is $\lceil \log_2 n \rceil$; hence the running time is n^c, which is a polynomial in n. As n is the length of the input 0^n, the running time of M' is polynomial in n.

input n
simulate M on input 0^n
accept if and only if M accepts
end

Figure 4.4   Deciding the "binary" version of a tally set

For the converse, let M be a machine that accepts the tally version of A in polynomial time. Consider the machine M' of Figure 4.4. By use of an argument analogous to the preceding, it is easily seen that M accepts A in time $2^{c \cdot |n|}$ for some constant c.

Part (b) is identical, using nondeterministic machines M and M'.   □

The next result, which is now a simple corollary to the previous proposition, establishes a relationship between the $P \overset{?}{=} NP$ question and the similar question $DEXT \overset{?}{=} NEXT$.

Theorem 4.19 $DEXT \neq NEXT$ if and only if there are tally sets in $NP - P$.

Proof.   If $DEXT \neq NEXT$ then there is a set A in $NEXT$ which is not in $DEXT$. By the previous proposition, its tally version is in NP but not in P.

Conversely, let T be a tally set in NP not in P. Let A be the following set: $A = \{n \mid 0^n \in T\}$. We have that $T = tally(A)$. Thus, by the previous proposition, A is in $NEXT$ but is not in $DEXT$, and therefore these two classes differ.   □

Now Corollary 3.33 follows here again.

Corollary 4.20   *If $P = NP$ then $DEXT = NEXT$.*

There is a similarity between tally sets and sparse sets. We proved in Lemma 1.10 that every tally set is sparse. However, although not every sparse set is a tally set, we can state the following relationship:

Theorem 4.21   *For every sparse set S there is a tally set T such that S is in $P(T)$. Furthermore, if S is in NP then T is in NP too.*

Proof.   Let S be given. Without loss of generality we assume that the empty word is not in S; otherwise it is treated by table look-up. For each length n, order lexicographically all the words in S of length up to n. Let $y_{n,j}$ be the jth word of length n in S in this lexicographical order.

Recall from Definition 1.8 that the census function $C_S(n)$ gives us the number of words in S of length up to n. As S is sparse, there is a polynomial p such that $C_S(n) \leq p(n)$.

Consider the tally set T accepted by the nondeterministic algorithm of Figure 4.6, and let us discuss three cases. If t is such that the value of k is larger than the census, there is no way of finding the k required words in S, and the algorithm will reject. If the value of k is smaller than the census, there are several ways of choosing the words, and the behavior is not predictable. But, if the value of k is exactly the census, then the only accepting computation must have guessed exactly the words in S up to length n, and therefore the test is trustworthy.

We will prove that T fulfills the conditions stated in the theorem. We prove first that S is in $P(T)$. Consider the oracle machine of Figure 4.5. Note that all the loops are polynomially bounded. The point of that algorithm is that T itself provides the correct value of the census; we use here the fact that $y_{n,1}$ is not the empty word. Once it is known, T provides any desired individual bit of any word in S.

```
input y
n := |y|
k := p(n)
while k ≥ 0 do
        if 0⟨n,k,1,1,0⟩ ∈ T or 0⟨n,k,1,1,1⟩ ∈ T
            then exit the while loop
comment: k is the correct value of C_S(n)
            else k := k − 1
comment: try other census
end while
if k = 0 then reject
comment: S empty up to this length
for j := 1 to k do
comment: construct y_{n,j}
    z := λ
    for i := 1 to n do
            if 0⟨n,k,i,j,0⟩ ∈ T then z := z0
            else if 0⟨n,k,i,j,1⟩ ∈ T then z := z1
            else exit the for loop
        if y = z then accept
end for
reject
comment: y is not among the y_{n,j}
end
```

Figure 4.5   Deciding a sparse set with a tally oracle set

The algorithm constructs each of the words $y_{n,j}$ in S and compares with the input y. In this way it accepts S with the help of the oracle T, in polynomial time.

It remains to show that if S is in NP then T is in NP. Let M be a nondeterministic polynomial time machine which accepts S, and substitute calls to M for the tests of membership in S in the algorithm of Figure 4.6. Observe that the length of the input is t, and that all the steps in that algorithm can be performed in time polynomial in t.   □

```
input x
if x is not of the form 0^t then reject
let t = ⟨n, k, i, j, b⟩
guess k words y_{n,m} of length at most n
check that they have been guessed in lexicographic order
for each of y_{n,m} do
      nondeterministically check that y_{n,m} is in S
check that the i^{th} bit of y_{n,j} is b
if all that holds then accept
end
```

Figure 4.6   A nondeterministic algorithm for a tally set

Our next corollary is an important result, based on the previous theorem. It shows that Theorem 4.19 holds not only for the tally sets but more generally for the sparse sets.

Corollary 4.22 $DEXT \neq NEXT$ if and only if there exists a sparse set in $NP - P$.

Proof.   If $DEXT \neq NEXT$ then by Theorem 4.19 there is a tally, hence a sparse set in $NP - P$.

Conversely, assume that there is a sparse set S in $NP - P$ and consider the tally set T furnished by Theorem 4.21. We have that T is also in NP. But T cannot be in P. Indeed, since S is in $P(T)$ and P is closed under polynomial time T-reducibility, if T is in P then S must be also in P. Thus, T is in $NP - P$, and by Theorem 4.19 we obtain that $DEXT \neq NEXT$.   □

4.4 Strong Nondeterministic Polynomial Time Reducibility

Let us now introduce another type of reducibility: the strong nondeterministic reducibility. It can be defined in terms of the many-one reducibility or the Turing reducibility. We present here the Turing case, leaving the other for the exercises (Exercise 23).

Definition 4.23 *A set A is strong nondeterministic T-reducible to B (denoted $A \leq^{SN} B$) if and only if $A \in NP(B) \cap co\text{-}NP(B)$.*

This reducibility will also be called SN-reducibility. A machine model that defines this reducibility is presented in Exercise 25. It has a particular interest: it is transitive, while the usual nondeterministic reducibility is not. The transitivity will follow from the next theorem, which will be used later on:

Theorem 4.24 *For all sets A, B, and C, if $A \in NP(B)$ and $B \leq^{SN} C$ then $A \in NP(C)$.*

Proof. Let $A \in NP(B)$, and let M_1 be a polynomial time nondeterministic oracle Turing machine such that $L(M_1, B) = A$.

Since $B \leq^{SN} C$, there are two nondeterministic polynomial time oracle machines M_2 and M_3 such that $B = L(M_2, C)$ and $\overline{B} = L(M_3, C)$.

Let us consider the nondeterministic polynomial machine M such that on instance x, M simulates the computation of M_1 on x; whenever M_1 enters a QUERY state, M uses the subroutine of Figure 4.7 to answer the query.

```
let w be the queried word
nondeterministically guess whether the answer is YES or NO
if the guess is YES then
      simulate M₂ on w using oracle C
      if M₂ accepts w under oracle C
            then continue M's computation with the YES answer
            else stop M's computation in a rejecting state
if the guess is NO then
      simulate M₃ on w using oracle C
      if M₃ accepts w under oracle C
            then continue M's computation with the NO answer
            else stop M's computation in a rejecting state
end
```

Figure 4.7 Answering queries nondeterministically

Therefore $A = L(M, C)$. □

As a corollary we obtain the transitivity of the SN-reducibility.

Corollary 4.25 *If $A \leq^{SN} B$ and $B \leq^{SN} C$ then $A \leq^{SN} C$.*

Proof. If $A \leq^{SN} B$ then both $A \in NP(B)$ and $\overline{A} \in NP(B)$. By the previous theorem, this fact, together with the fact that $B \leq^{SN} C$, implies that $A \in NP(C)$ and that $\overline{A} \in NP(C)$. Therefore $A \in NP(C) \cap$ co-$NP(C)$. □

One of the most important properties of the SN polynomial time reducibilities is that they determine a class of complete sets in NP which are not in co-NP unless $NP =$ co-NP. Therefore these reducibilities can be used to demonstrate the intractability of a problem (i.e. that the problem is not in P) based on the hypothesis that $NP \neq$ co-NP. The definition of completeness is identical to other completeness definitions:

Definition 4.26 *Given a class $\mathcal{C}$,*

1. *A set A is $\mathcal{C}$-SN-hard, or SN-hard for $\mathcal{C}$, if and only if for any set $B \in \mathcal{C}$, $B \leq^{SN} A$.*
2. *A set A is $\mathcal{C}$-SN-complete, or SN-complete for $\mathcal{C}$, if and only if it is $\mathcal{C}$-SN-hard and $A \in \mathcal{C}$.*

The following properties of the SN-reducibility and of the SN-complete sets can be proved.

Proposition 4.27
(a) $\leq^{SN}$ *is a preorder.*
(b) *For every A, B, $A \leq^{SN} A \oplus B$ and $B \leq^{SN} A \oplus B$.*
(c) *For every A, B, C, if $A \leq^{SN} C$ and $B \leq^{SN} C$ then $A \oplus B \leq^{SN} C$.*
(d) *$A \leq^{SN} B$ if and only if $\overline{A} \leq^{SN} B$, if and only if $A \leq^{SN} \overline{B}$, if and only if $\overline{A} \leq^{SN} \overline{B}$.*
(e) *If $A \in NP \cap$ co-NP then, for any other set B, $A \leq^{SN} B$.*
(f) *If A is $\mathcal{C}$-SN-hard and $A \leq^{SN} B$ then B is $\mathcal{C}$-SN-hard.*
(g) *If A is $\mathcal{C}$-SN-complete, $B \in \mathcal{C}$, and $A \leq^{SN} B$, then B is $\mathcal{C}$-SN-complete.*
(h) *$NP \cap$ co-NP and $PSPACE$ are closed under this reducibility.*
(i) *If A is NP-SN-complete or co-NP-SN-complete, then $A \in NP \cap$ co-NP if and only if $NP =$ co-NP.*

The proof is left once more as an exercise (see Exercise 28).

To end this section, observe that this reducibility yields an equivalence relation, whose classes are the SN-degrees, and a partial order among the SN-degrees. The argument for obtaining this equivalence is the same as before. The equivalence relation is denoted $\equiv^{SN}$. Notice that the infimum of this partial order is the class $NP \cap$ co-NP.

4.5 Self-Reducibility

The last property we shall talk about in this chapter is self-reducibility. This is a very important property which will be used later on.

Informally, a set A is self-reducible if the membership problem of an element is "reducible" to the membership problems of smaller elements.

Definition 4.28 *A set A is self-reducible if and only if there exists a deterministic polynomial time oracle machine M, such that the following holds:*

1. $A = L(M, A)$.
2. *On each input of length n, M queries the oracle only about strings of length at most $n - 1$.*

Notice that condition 1. can be interpreted as saying that A is a "fixed point" of M. In fact, using condition 2., it is not difficult to prove that such fixed points are unique (see Exercise 30).

Some important sets are self-reducible. In particular, the self-reducibility of SAT will be important in Chapter 5 of this volume, and in other chapters of Volume II.

Theorem 4.29 *The set* SAT *is self-reducible.*

Proof. We present a deterministic procedure describing the computation of a machine M as required by Definition 4.28. It will accept SAT when it is given SAT as oracle.

The procedure is presented in Figure 4.8.

---

input F
if F does not have variables then
    simplify F
    accept if and only if F simplifies to "true"
else let x be the variable in F with smallest index
if $F|_{x:=0}$ is in the oracle
or $F|_{x:=1}$ is in the oracle
    then accept
    else reject
end

Figure 4.8 Self-reducibility algorithm for SAT

It is not difficult to encode the formulas in such a way that the simplifications $F|_{x:=0}$ and $F|_{x:=1}$ are both shorter than F. □

The set QBF is also self-reducible (see Exercise 31). This fact will be used later, too. A polynomial space upper bound on the space complexity of every self-reducible set is presented in Exercise 32.

4.6 Exercises

1. Prove Proposition 4.3.

2. Prove that in the semilattice of T-degrees there is no maximum polynomial T-degree.

3. Show that for every oracle A, if $P(A) = PSPACE(A)$ then $PF(A) = PSPACEF(A)$.

4. •Show that NP is closed under $\leq_T$ if and only if NP is closed under complementation.

5. Define a space bounded analogue of the T-reducibility. Discuss the validity of the statement of Theorem 3.38 for this reducibility.

6. Prove Proposition 4.5.

7. Prove Proposition 4.11.

8. Show that $A \leq^{NP} B$ if and only if $A \leq_m K(B)$.

9. In Volume II it will be proved that the nondeterministic polynomial time Turing reducibility is not transitive. Which ones among the remaining statements of Proposition 4.3 hold for the nondeterministic polynomial time Turing reducibility? Prove your answer.

10. The *polynomial time truth-table reducibility* is defined as follows: $A \leq_{tt} B$ if and only if $A \leq_T B$ via a machine which writes down in a separate tape all the queries to be made during the computation before the first word is queried. Show that a set A is in $P(T)$ for a tally set T if and only if $A \leq_{tt} T'$ for some tally set T'.

11. The *polynomial time bounded-truth-table reducibility* is defined as follows: $A \leq_{btt} B$ if and only if $A \leq_{tt} B$ via a machine that on each input queries the oracle at most k times, where k is a constant independent of the inputs. Let A be any set. Show that the class of sets which are btt-reducible to A is the boolean closure of the class of sets which are m-reducible to A.

12. Show that for any set A, $A \leq_{btt} T$ for a tally set T if and only if $A \leq_m T'$ for some tally set T'.

13. The *conjunctive polynomial time reducibility* is defined as follows: $A \leq_c B$ if and only if $A \leq_{tt} B$ via a machine that accepts the input exactly when all the queries have been answered positively by the oracle. Show that NP is closed under this reducibility, i.e. if $B \in NP$ and $A \leq_c B$ then $A \in NP$.

14. The *disjunctive polynomial time reducibility* is defined as follows: $A \leq_d B$ if and only if $A \leq_{tt} B$ via a machine that accepts the input exactly when at least one of the queries has been answered positively by the oracle. Show that $A \leq_d B$ if and only if $\overline{A} \leq_c \overline{B}$. Deduce from this fact and the previous exercise that if $B \in$ co-NP and $A \leq_d B$ then $A \in$ co-NP.

15. An oracle machine M is *positive* if and only if $A \subseteq B \Rightarrow L(M, A) \subseteq L(M, B)$. The *positive Turing reducibility* is defined as follows: $A \leq_{pos} B$ if and only if $A = L(M, B)$ where M is a positive oracle machine. Show that NP is closed under $\leq_{pos}$.

16. Using the concept of the previous exercise, define the *positive truth-table reducibility* and the *positive bounded-truth-table reducibility*. Use the previous exercise to improve Exercises 13 and 14, showing that both *NP* and co-*NP* are closed under both $\leq_c$ and $\leq_d$.

17. State and prove an analogue to Proposition 4.11 for classes of the form $P(\mathcal{C})$ or $NP(\mathcal{C})$, instead of $P(A)$ or $NP(A)$. Here $\mathcal{C}$ is any complexity class.

18. Prove Part (b) of Proposition 4. 18.

19. Prove Theorem 4.16. Define also a deterministic version $K_d(A)$ of the $K(A)$ operator, using deterministic machines, and prove that $K_d(A)$ is $P(A)$-complete. Deduce that a class closed under $\leq_T$ has a T-complete set if and only if it has an m-complete set.

20. For any set A, define

$$KS(A) = \{\langle M, x, 1^k\rangle \mid \text{the deterministic machine } M \text{ accepts } x$$

relative to A using an amount of space bounded by k }

Prove that $KS(A)$ is $PSPACE(A)$-m-complete.

21. For any set A, define

$$KE(A) = \{\langle M, x, k\rangle \mid \text{the deterministic machine } M \text{ accepts } x$$

relative to A in at most k steps }

where k is written in binary. Prove that $KE(A)$ is $EXPTIME(A)$-m-complete.

22. Prove that if T is a tally set, then $DEXT(T) = DEXT$ implies that T belongs to P.

23. The *strong nondeterministic polynomial time many-one reducibility*, also called γ-*reducibility*, is defined as follows: $A \leq_m^{SN} B$ if and only if there is a polynomial time nondeterministic machine with output tape, such that, on every input x, at least one computation path outputs some result, and the output of each computation path is in B if and only if x is in A. Which of the statements of Proposition 4.3 hold for this reducibility? Prove your answer.

24. A *strong nondeterministic machine* is a machine with accepting and rejecting states, such that for each input x, either some computation path accepts x and no computation path rejects x, or some computation path rejects x and no computation path accepts x. Show that a set is accepted in polynomial time by a strong nondeterministic machine if and only if it is in $NP \cap$ co-*NP*.

25. Given an oracle A, a *strong nondeterministic oracle machine under A* is a machine with accepting and rejecting states, such that for each input x, either some computation path accepts x and no computation path

rejects x, or some computation path rejects x and no computation path accepts x. Show that a set B is accepted by a strong nondeterministic machine under A if and only if $B \leq^{SN} A$.

26. Show that there are nondeterministic oracle machines which are strong under some oracles but are not strong under other oracles.

27. What is the relationship between the strong nondeterministic polynomial time m-reducibility and the nondeterministic polynomial time m-reducibility? Answer the analogous question for Turing reducibilities, too.

28. Prove Proposition 4.27.

29. Show that the following are equivalent:

    (a) $B \leq^{SN} A$;
    (b) $K(B) \leq_m K(A)$;
    (c) $NP(B) \subseteq NP(A)$.

30. Prove that if M is the machine which witnesses the self-reducibility of a set A, and B is any set such that $B = L(M, B)$, then $A = B$.

31. Prove that QBF is self-reducible.

32. Prove that if a set A is self-reducible then A belongs to $PSPACE$.

33. • Robust Turing machines were defined in Definition 1.50. An oracle A *helps* a robust machine M if and only if for some polynomial p and for every input x, the running time of M on x with oracle A is at most $p(|x|)$. Define $P_{help}(A)$ as the class of sets $L(M, A)$ where M is robust and A helps M, $P_{help}(\mathcal{C})$ as the union of all the classes $P_{help}(A)$ for A in $\mathcal{C}$, and P_{help} as $P_{help}(\mathcal{P}(\Sigma^*))$. Prove:

    (a) For every set B, $P_{help}(B) \subseteq P(B)$.
    (b) If $A \in P(B)$ then $P_{help}(A) \subseteq P_{help}(B)$.
    (c) If $A \in P_{help}(B)$ then $P(A) \subseteq P_{help}(B)$.
    (d) $P_{help} = NP \cap$ co-NP.

34. Show that if A helps every set in $NP \cap$ co-NP then A is T-hard for $NP \cap$ co-NP.

35. Define NP_{help} as the class of sets which are accepted by nondeterministic robust oracle machines in polynomial time with the help of some oracle. Show that $NP_{help} = NP$.

4.7 Bibliographical Remarks

As mentioned in the remarks at the end of Chapter 3, the semilattice structure of the polynomial time degrees was studied by Ladner (1975a). This work studies both m and T degrees.

The relationship between the polynomial time many-one reducibility and the polynomial time Turing reducibility has established by itself a sound line of research, which we do not go into in this volume. We shall just mention some of the results for the interested reader. Ladner, Lynch and Selman (1975) have established the existence of a set A in $DEXT$ such that $A \not\leq_m \overline{A}$ and $A \leq_T \overline{A}$. We derive this result from the bi-immunity techniques in Volume II. In this same paper, the deterministic and nondeterministic versions of "truth-table", "bounded truth-table", "conjunctive", "disjunctive", and "positive" reducibilities are defined, and several results proved showing that they differ. We have taken from that work some questions presented in Exercises 13 to 16. Exercise 11 is from Köbler, Schöning, and Wagner (1987), and Exercises 10 and 12 are from Book and Ko (1988). Exercise 19 is from Ambos-Spies (1987).

The study of the relationship between complexity classes using the properties of tally and sparse sets was initiated by Book (1974b); from this last work comes Theorem 4.19. Hartmanis and his school have studied in depth the density properties of complexity classes, a taste of which is given by Theorem 4.21 and Corollary 4.22. This theorem is from Hartmanis (1983). In a later work, Hartmanis and Yesha (1984) show a more general version of the same theorem. In Volume II we shall present more results relating density properties.

The concept of strong nondeterministic polynomial time reducibility was fully developed by Long (1982), based on the concept of γ-reducibility due to Adleman and Manders (1977). Exercises about strong machines and SN-reducibility are from these papers, and also from Selman (1978). Long treats not only the Turing SN-reducibility, but also the many-one and the truth-table reducibilities, comparing them with other kinds of reducibilities.

The concept of self-reducibility appears in Schnorr (1976a). In this work, a collection of self-reducible sets is presented. The notion of self-reducible set was more explicitly used later in Meyer and Paterson (1979). Some of the results presented here and in the exercises are from Ko (1983) and Balcázar, Book, and Schöning (1986).

The concept of "helping" is due to Schöning (1985b). Exercises on P_{help} and NP_{help} are from this reference. Ko (1987) has continued the study of this subject.

5 Nonuniform Complexity

5.1 Introduction

The complexity measures and the classes introduced so far are intrinsically algorithmic. Our interest is in the amount of resources used by the algorithms solving the problems. But any finite set can be recognized in constant time and zero work space by a deterministic finite automaton; therefore, measuring the amount of resources is meaningless when considering only finite sets. The "intrinsically algorithmic" approach makes sense only when dealing with infinite sets.

We introduce in this chapter the "nonuniform" approach, which is a tool for dealing with finiteness. Instead of measuring *resources* used by algorithms accepting infinite sets, we measure *sizes* of algorithms accepting finite sets. Then we can associate with each infinite set a function describing the growing rate of the sizes of the algorithms for the initial segments of the set.

In order to establish a connection between these two approaches, we first introduce the concept of "advice functions". This unifying concept allows us to present the main definitions with a minimum of technicalities, and to enter gradually into the deeper study of the main nonuniform measures.

5.2 Classes Defined by Advice Functions

Consider any class of functions from $\mathbb{N}$ to Σ^*; call these functions *advice functions*. We are going to consider in this chapter complexity classes obtained by assuming that we have an "advice" function available that helps us in solving our problems, and measuring the amount of resources needed by our algorithms modulo the appropriate advice function.

The main characteristic of advice functions is that the value of the advice does not fully depend on the particular instance we want to decide, but just on its length. Hence, advice functions provide external information to the algorithms, as oracles do, but the information provided by an oracle may depend on the actual input, whereas the information provided by an advice function does not: just the length of the input matters.

We formalize this notion as follows.

Definition 5.1 *Let C be a class of sets, and let $\mathcal{F}$ be a class of advice functions. The class $C/\mathcal{F}$ is the class of all the sets B for which there exists a set $A \in C$ and a function $f \in \mathcal{F}$ such that $B = \{x \mid \langle x, f(|x|)\rangle \in A\}$.*

The classes obtained in this way are known as *nonuniform* classes. If C is a complexity class defined by a bound on the amount of some computational resource, $C/\mathcal{F}$ is the class of all sets B such that some function of $\mathcal{F}$ provides enough additional information to solve B within the bounds specified by the class. Usually the family of functions $\mathcal{F}$ is defined by some bound on the lengths of the values in terms of the argument. Occasionally we will speak of the resources needed to compute an advice function; in this case, the argument is assumed to be given in unary notation, so that advice functions are functions from $\{0\}^*$ into Σ^*.

Notice that, in this definition, the nonuniform information is provided by the advice function, and that it depends only on the length of the input, being the same for all the input words of the same length, as indicated above.

Let us turn to some examples; in this chapter we will study in depth some of them, which will be used later.

Example 5.2 Let *log* denote the family of functions f from $\mathbb{N}$ to Σ^* such that for each n, and for some positive constant c, $|f(n)| \leq c \cdot \log n$. Consider the complexity class P. The nonuniform class P/log is then the class of sets $B = \{x \mid \langle x, f(|x|)\rangle \in A\}$, where $A \in P$ and $f \in log$. In other words, a set B is in the class P/log if for any w in Σ^* we are able to decide if w is in B with a deterministic machine working in polynomial time, with the help of an advice which has a length logarithmic in the size of the input.

Example 5.3 Let *poly* denote the family of functions f from $\mathbb{N}$ to Σ^* such that for some polynomial p, $|f(n)| \leq p(n)$ for each n. If this class is used instead of *log* in the previous example, we obtain $P/poly$, which is the class of sets $B = \{x \mid \langle x, f(|x|)\rangle \in A\}$, where $A \in P$ and for some polynomial p and every n, $|f(n)| \leq p(n)$.

Example 5.4 Consider the class *poly* as in the previous example, and the complexity class $PSPACE$. The nonuniform class $PSPACE/poly$ is then the class of sets $B = \{x \mid \langle x, f(|x|)\rangle \in A\}$, where $A \in PSPACE$ and $f \in poly$.

Some other examples are presented in the Exercises. We will characterize later on some of these classes, and show several properties which will be used in other parts of this book. Some prior results about boolean circuits and their relationship with the computations of Turing machines are needed.

5.3 Boolean Circuit Complexity

Let us introduce our first type of nonuniform complexity measure: the boolean circuit complexity. We fix the following family of boolean operations, called *basis:* the constants—i.e. 0-ary operations—"true" and "false" (which are sometimes also denoted 1 and 0); the unary operation of negation, denoted $\neg$; and the binary operations $\wedge$ and $\vee$. We use these basic functions to compute other n-ary boolean functions.

Definition 5.5   *Given a set $\Delta = \{x_1, x_2, \ldots, x_n\}$ of n boolean variables, a computation chain over Δ is a sequence $g_1, g_2, \ldots, g_k$, in which each g_j is either an element of Δ, or a boolean constant, or a pair $(\neg, g_l)$ for $1 \leq l < j$, or a triple $(\wedge, g_l, g_m)$ for $1 \leq l, m < j$, or a triple $(\vee, g_l, g_m)$ for $1 \leq l, m < j$.*

The symbol g for the elements of the computation chain stands for "gate". However, for historical reasons, not all the elements of a computation chain are considered gates:

Definition 5.6   *The elements of a computation chain over Δ which consist of a variable of Δ or of a boolean constant are called source elements. The remaining elements are called computation elements or gates. The inputs to a gate g_i are the (one or two) smaller elements of the computation chain which appear in g_i.*

With each element g_i of a computation chain over $\Delta = \{x_1, \ldots, x_n\}$, we can associate an n-ary function result(g_i) which represents the boolean value computed at element g_i. It is defined as follows.

Definition 5.7

$$
result(g_i) = \begin{cases}
g_i & \text{if } g_i \text{ is a source} \\
\neg result(g_l) & \text{if } g_i = (\neg, g_l) \\
result(g_l) \wedge result(g_m) & \text{if } g_i = (\wedge, g_l, g_m) \\
result(g_l) \vee result(g_m) & \text{if } g_i = (\vee, g_l, g_m)
\end{cases}
$$

It should be noticed that in this definition the symbols of boolean operation denote the corresponding operation, so that "result$(g_l) \wedge$ result(g_m)" denotes the boolean value "true" or "false" obtained by applying the conjunction operation to the boolean values obtained by the n-ary boolean function "result" applied to g_l and g_m.

Representing computation chains by acyclic graphs we obtain circuits:

Definition 5.8   *A boolean circuit, also called a combinational machine or just a circuit, is an acyclic graph representation of a computation chain, which is constructed by associating to each step g_i of the chain a node labeled with the variable, constant, or operation present in g_i, and joining node g_i to node g_j by a directed edge if g_i is an input to g_j.*

For each circuit, there exists an enumeration of its gates which coincides with the enumeration of the elements in the computation chain represented by the circuit. Therefore each circuit can be identified with the computation chain it represents.

Definition 5.9 *Given a boolean function f from $\{0,1\}^n$ to $\{0,1\}^m$, which can be expressed as an m-tuple $f = (f_1, \ldots, f_m)$ of boolean functions from $\{0,1\}^n$ to $\{0,1\}$, we say that a circuit $C = (g_1, \ldots, g_k)$ computes f if for every r, $1 \leq r \leq m$, there exists an s, $1 \leq s \leq k$, such that $f_r = result(g_s)$.*

Thus, a circuit computes such a function if each of the components of the function is the result of some gate (or source) of the circuit. Let us consider an example.

Example 5.10 Consider the boolean function f from $\{0,1\}^4$ to $\{0,1\}^3$ defined by the triple (f_1, f_2, f_3), where

$$f_1(x_1, x_2, x_3, x_4) = x_1 \wedge (x_2 \vee x_3)$$

$$f_2(x_1, x_2, x_3, x_4) = \neg(x_2 \vee x_3) \vee (x_4 \wedge 0)$$

$$f_3(x_1, x_2, x_3, x_4) = f_2 \wedge 0$$

Figure 5.1 shows an example of a circuit which computes the boolean function f just described.

In many cases, the boolean function we expect a circuit to compute has only one component, i.e. it is a function f from $\{0,1\}^n$ into $\{0,1\}$. In this case, the gate g_i for which $result(g_i) = f$ is called the *output* of the circuit. Unless otherwise stated, we follow the convention that the output gate of a circuit $C = (g_1, g_2, \ldots, g_k)$ is the gate g_k.

Definition 5.11 *Given a circuit C, a standard encoding of C is a string of 4-tuples of the form (g, b, g_l, g_r), where g represents the gate number, b is the boolean operation performed by the gate, g_l is the number of the gate which provides the left input to g, and g_r is the number of the gate, variable, or constant which provides the right input to g. For negation gates, g_r is some null code.*

This string can be encoded in a standard way over $\{0,1\}$, as indicated in Chapter 1. Observe that, on any alphabet having more than one symbol, it is easy to construct encodings of circuits whose length is polynomial in the number of gates.

Example 5.12 For the circuit of Figure 5.1, one of its standard encodings is the following:

$$(5, \vee, 2, 3)(6, \wedge, 4, F)(7, \wedge, 1, 5)(8, \neg, 5)(9, \vee, 8, 6)(10, \wedge, 9, F)$$

which can be encoded over any other alphabet in the canonical way.

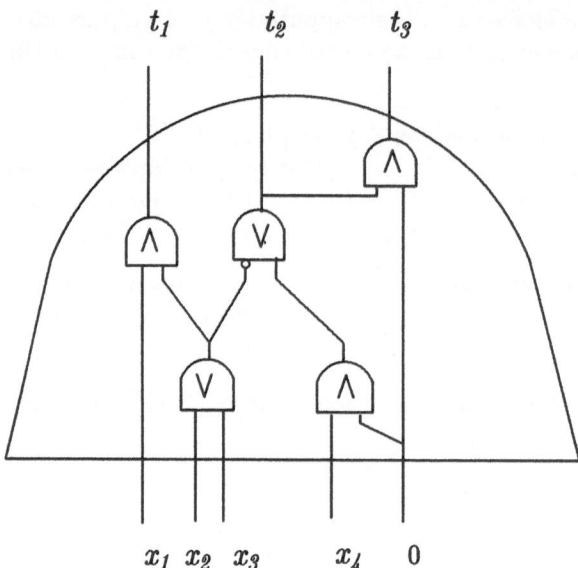

t_1 t_2 t_3

x_1 x_2 x_3 x_4 0

Figure 5.1   A boolean circuit

Let us define the two kinds of nonuniform complexity measures we shall be dealing with.

Definition 5.13   *The cost or size of a circuit is the number of gates it has. Given a boolean function f, its boolean cost $c(f)$ is the size of the smallest circuit computing it.*

Definition 5.14   *The depth of a circuit is the length of the longest path in the graph of the circuit. Given a boolean function f, denote by $d(f)$ its boolean depth, which is the depth of the minimal depth circuit computing f.*

Let us present some examples of these two types of boolean complexity.

Example 5.15   We start by computing the dot product of two boolean vectors. Given two vectors of dimension n, vector $\vec{x} = (x_1, \ldots x_n)$ and vector $\vec{y} = (y_1, \ldots y_n)$, their *dot product* is the function from $\{0,1\}^{2n}$ to $\{0,1\}$ defined by

$$\mathrm{dot}(\vec{x}, \vec{y}) = \mathrm{dot}(x_1, \ldots, x_n, y_1, \ldots, y_n) = \bigvee_{i=1}^{n} (x_i \wedge y_i)$$

The circuit in Figure 5.2 computes this function. This circuit shows that $d(\mathrm{dot}(\vec{x}, \vec{y})) \leq \lceil \log n \rceil + 1$ and that $c(\mathrm{dot}(\vec{x}, \vec{y}))$ is $O(n)$.

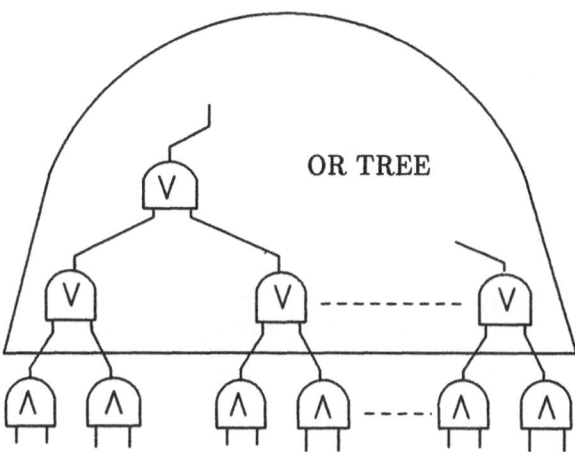

Figure 5.2   Computing a dot product

Example 5.16   Let us turn now to the matrix product of two square boolean matrices $A = [a_{i,j}]$ and $B = [b_{i,j}]$ of dimension n. If $C = [c_{i,j}]$ is the matrix product $A \cdot B$, then $c_{i,j} = \text{dot}(A_i, B_j)$ where A_i is the row vector $(a_{i,1}, \ldots, a_{i,n}), 1 \leq i \leq n$, and B_i is the column vector $(b_{1,i}, \ldots, b_{n,i}), 1 \leq i \leq n$.

Note that all $c_{i,j}$ can be calculated in parallel by many circuits computing a different dot product each, having no interconnections among them. Each input is connected to n different gates. Such a circuit has depth $\lceil \log n \rceil + 1$.

Example 5.17   Our next example, which will be used later on, deals with the transitive closure of a boolean square matrix. Given a matrix $A = [a_{i,j}]$, $1 \leq i, j \leq n$, its *reflexive transitive closure* is defined as:

$$A^* = [a_{i,j}^*] = \bigvee_{i \geq 0} A^i$$

where A^i is the ith product of A by itself. By definition, $A^0 = I$, the identity matrix.

It is easy to see that $a_{i,j}^* = 1$ if and only if either $i = j$, or there exists a sequence $k_1, \ldots, k_l$, with $l \geq 1$ and each k_i between 1 and n, such that each of the entries $a_{i,k_1}, a_{k_1,k_2}, \ldots, a_{k_l,j}$ of A is 1. Moreover, all such $k_1, \ldots, k_l$ can be assumed different, and therefore we can take $0 \leq l \leq n$. Hence, the sum can be replaced by $A^* = (A + I)^n$.

We can evaluate $(A + I)^n$ using the scheme of Figure 5.3. Therefore the transitive closure can be computed in a depth bounded by the square of the logarithm of the dimension of the matrix.

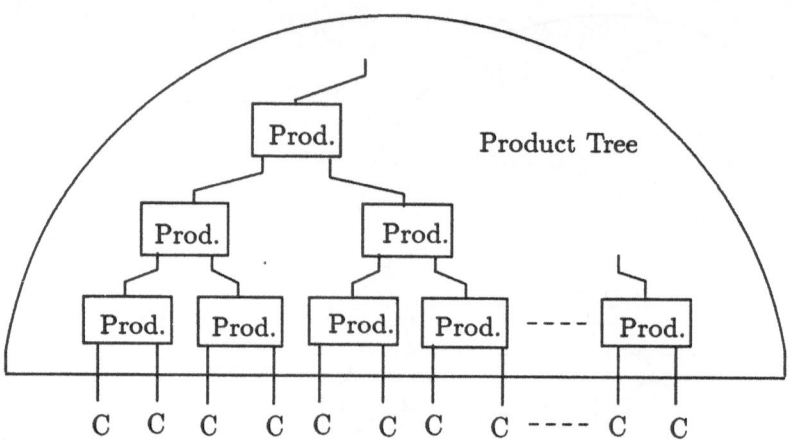

Figure 5.3   Computing a transitive closure; C stands for $(A + I)$

5.4 Turing Machines and Boolean Circuits

As indicated above, boolean circuits have a fixed number of input bits. Hence, when faced with an infinite set, we should use a different circuit for each length. Throughout this section, infinite sets are subsets of $\{0,1\}^*$.

Definition 5.18   *Let $A \subseteq \{0,1\}^*$. For any given n, we define the circuit complexity of A at length n, $c_A(n)$, as the boolean cost of the characteristic function of $A \cap \{0,1\}^n$; and we define the depth complexity of A at length n, $d_A(n)$, as the boolean depth of the same function. Observe that as functions of n, both c_A and d_A only depend on A. Thus, we call the function c_A the circuit complexity of A, and we call the function d_A the depth complexity of A.*

In this section we shall relate circuit complexity and depth complexity to time and space complexity of Turing machines. We start by introducing a codification of the configurations and the computations of Turing machines in terms of boolean functions.

Let us consider a one tape deterministic Turing machine M, with tape alphabet $\Sigma = \{a_1, a_2, \ldots a_n\}$, set of states $Q = \{q_1, \ldots q_m\}$, and transition function δ. Let $r = \lceil \log |Q| \rceil + 1$ and $r' = \lceil \log |\Sigma| \rceil + 1$. Encode each state of Q in binary form as a sequence of r bits, excluding the sequence 0^r. Similarly, encode each symbol of Σ in binary by a sequence of r' bits, excluding again the sequence $0^{r'}$.

At each instant t of a computation of M, we associate with each cell u of M a representation of its parameters: the contents of the cell, whether the head is on or off the cell, and in the case when the head is on the cell, the state of M at that moment. This representation is given by a pair $u^t = \langle s^t, x^t \rangle$, where s is called the *state part*, and is defined by

$$s^t = \begin{cases} 0^r, & \text{if the head of } M \text{ is not on } u \text{ at time } t \\ q_i, & \text{if at time } t, M \text{ is on } u \text{ and in state } q_i \end{cases}$$

while x^t is the *symbol part*, which gives the contents of cell u at instant t.

During a computation of M, the change of the contents of a cell depends only on the contents of the cell itself and on the contents of the cells at its right and at its left. Consider the representations of three consecutive cells at time t: u_1^t, u_2^t, and u_3^t. The change of representation from u_2^t to u_2^{t+1} can be due to three causes:

1. The tape head is on cell u_1 and it moves right. In this case u_2 contains a new state and the symbol part remains unchanged.
2. The tape head is on cell u_3 and it moves to the left. This case is similar to case 1.
3. The tape head is on u_2. Then, depending on the transition function, the cell may change either the state, the symbol part, or both.

To describe such a cell change, we introduce two boolean functions, describing respectively the changes of the state part and of the symbol part; we call them "state" and "symbol".

The function state:$\{0,1\}^{r+r'} \rightarrow \{0,1\}^{3 \cdot r}$ will be defined so that it encodes simultaneously the move of the head and the change of state.

We define:

$$\text{state}(q, x) = \begin{cases} (q', 0^r, 0^r) & \text{if the head moves to the left} \\ (0^r, q', 0^r) & \text{if the head does not move} \\ (0^r, 0^r, q') & \text{if the head moves to the right} \end{cases}$$

where q' is the next state as indicated by the transition function $\delta(q, x)$ of M. We also define state$(0^r, x) = (0^r, 0^r, 0^r)$. Denote by l-state, m-state, and r-state the functions resulting from extracting the left, middle, and right part, respectively, of the value state(q, x). Note that the state part of the description of cell u_2 at time $t + 1$ is the OR of the values of l-state(u_3), m-state(u_2), and r-state(u_1) at time t, and that each of these values corresponds to one of the three cases described above.

The function symbol:$\{0,1\}^{r+r'} \rightarrow \{0,1\}^{r'}$ computes the next symbol to be printed on a cell, and is defined as follows: symbol$(0^r, x) = x$, because only the presence of the head may change the symbol; and symbol(q, x) is the symbol printed over x by M in state q, as indicated by $\delta(q, x)$.

Notice that the functions "state" and "symbol" depend only on the transition function of M. They depend neither on the length of the input word, nor on the run time of M nor the space used by the machine, hence they can be computed by fixed circuits of constant size and depth.

The next theorem relates size complexity and the time of a Turing machine computation.

Theorem 5.19 *If A is accepted by a deterministic one tape Turing machine in time $T(n)$, then $c_A(n) = O(T^2(n))$.*

Proof. Let M be a one tape machine that accepts A in time $T(n)$. Assume that M has only one final (i.e. accepting) state q_f and that when M enters state q_f, it keeps looping forever in the same state. Therefore, M will be in state q_f at time $T(n)$, on accepted inputs.

Now we shall construct a boolean circuit that simulates the evolution of the tape contents of machine M on input w during exactly $T(n)$ steps. To accept a word, the circuit will test whether the state of M is q_f after $T(n)$ steps. We use the previously developed formalism to compute the contents of the cells and the state of M at time $t + 1$, from the contents of the cells and the state of M at time t. The whole circuit will have the form indicated in Figure 5.4, where each box represents a full copy of the circuit computing the state and the symbol part of a cell.

The box in the ith column and t^{th} row, $0 \le i, t \le T(n)$, computes the contents of the ith tape cell at time t (the symbol part), together with an indication of whether the head is reading cell i at time t; furthermore, if the head is present in this cell, it computes the state q of M at time t (the state part). The state part of u_j at time $t + 1$ is computed by OR-ing the results of the *l-state* and *r-state* functions at neighbour cells, u_{i-1} and u_{i+1} respectively, at time t, and the result of the *m-state* function at the same cell u_i at time t. The symbol part is computed directly by the function $symbol(q, x)$. The "transmission" of the information about the results of these functions is represented in the figure by an arrow between the boxes.

If M accepts an input w, then M reaches state q_f in at most $T(n)$ steps, so the tape head of M scans at most $T(n) + 1$ cells. Therefore the circuit has $T(n) + 1$ columns, each corresponding to a cell.

Each row corresponds to a configuration of the computation. The circuit has $T(n)$ rows. Let q_1 be the initial state of M, x_i the ith symbol of the input, and B the blank tape symbol. Then the inputs to the first row are defined by $u_0 = (q_1, x_1)$, $u_i = (0^r, x_i)$ for $1 \le i \le n$, and $u_i = (0^r, B)$ for other values of i.

At the last row, exactly one of the boxes will contain the state in which M halted at time $T(n)$. The remaining boxes do not contain any information on the *m*-state function, i.e. its value is 0^r. So the circuit can find the state by computing the bitwise OR of the results of the function *m*-state, and compare it with the known value of q_f by means of an "equivalence" circuit which consists of AND-ing the results of a bitwise equality. The result of this operation is the output of the circuit.

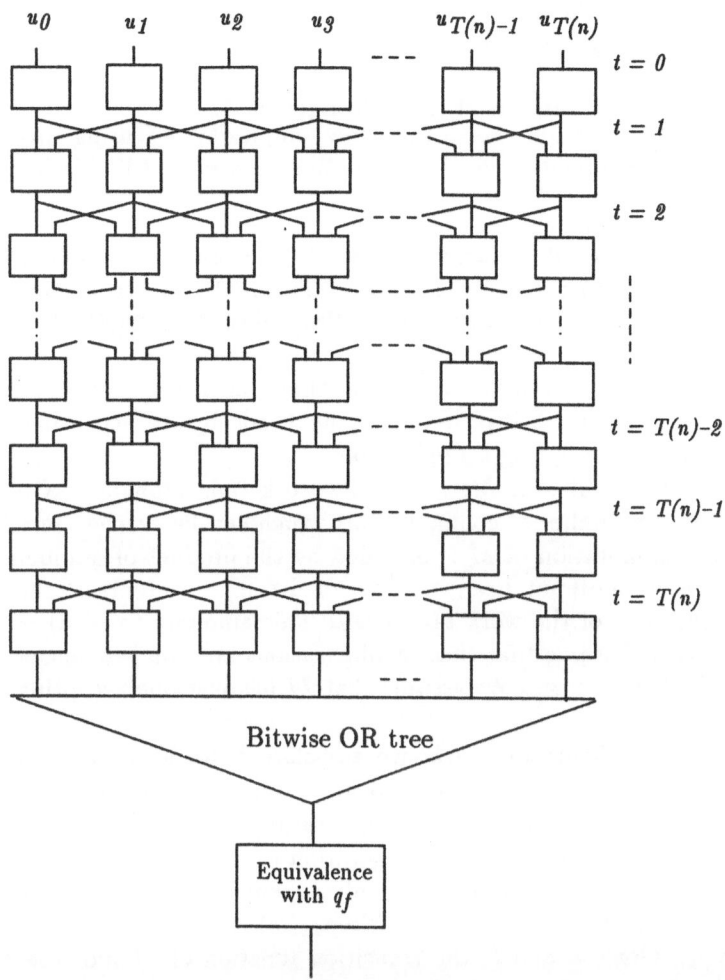

Figure 5.4   A circuit simulating a Turing machine

Each box in the network simulating the computation has cost $O(1)$, the bitwise OR can be computed by a tree of OR gates with cost $O(T(n))$, and the final equality circuit has cost $O(1)$. Therefore, the total cost of the whole circuit is bounded by $O(T^2(n))$.   □

Note that, instead of extracting the state, we can substitute the last part of the circuit by another circuit that extracts the contents of the tape. Thus, it follows that this theorem also applies to computing functions.

Corollary 5.20 (to the proof)   *If the function f is computed by a deterministic Turing machine with one tape within time $T(n)$, then the restriction of f to $\{0,1\}^n$ can be computed by a circuit of size $O(T^2(n))$.*

The formalization of this argument is left to Exercise 4.

Theorem 5.19 relates circuit size to the time complexity of Turing machines. The next theorem relates the circuit depth to space used by Turing machines.

Theorem 5.21   *Let M be a nondeterministic Turing machine with one work tape, a read-only input tape, and input tape alphabet $\{0,1\}$. If, for any $S(n) \geq \log n$, M accepts the set A within space $S(n)$, then $d_A = O(S^2(n))$.*

Proof. Without loss of generality, we assume that, on accepted inputs, M erases the work tape and brings left the tape heads before accepting, so that there is a unique accepting configuration.

Let Q be the set of states of M, and Σ its work tape alphabet. On inputs of length n, the number of configurations which can be reached in a shortest accepting computation of M is bounded by the product of number of states, positions of the input head, number of different contents of the work tape, and positions of the work tape head. This amounts to $N(n) = |Q| \cdot n \cdot |\Sigma|^{S(n)} \cdot S(n)$. Observe that these configurations are independent of the actual input $w = x_1 x_2 \ldots x_n$. We assume that M has just one accepting configuration.

To every pair of configurations a, b, we associate a boolean function $\text{Step}_{a,b} : \{0,1\}^n \rightarrow \{0,1\}$ such that $\text{Step}_{a,b}(w) = 1$ if and only if, when computing on input w, M can move from configuration a to configuration b in only one step. Note that if, while in configuration a, the input tape head is on cell k, then the fact that $\text{Step}_{a,b}(w) = 1$ only depends on the kth symbol x_k of the input w. In fact, for each a and b, $\text{Step}_{a,b}(w)$ is either 0, 1, x_k, or the negation of x_k. Given a and b, the transition function of M indicates which one of these four possibilities is the case. Thus, each of the functions $\text{Step}_{a,b}$ can be computed by a circuit of depth at most 1.

For each fixed input w, the values of $\text{Step}_{a,b}(w)$ form a $N(n) \times N(n)$ boolean matrix Step describing the possible transitions between configurations of M. Let Init be the initial configuration of M on w, and let Acc be the unique accepting configuration of M. We know that M accepts w if and only if there is a sequence of legal movements among configurations, starting at Init and ending at Acc. By the definition of the transitive closure of a boolean matrix, M accepts w if and only if there is a 1 in the entry $\text{Step}^*_{\text{Init,Acc}}$ in the transitive closure Step* of the matrix Step corresponding to input w.

In Example 5.17 of the previous section, we proved that transitive closure can be computed by a circuit with depth bounded by the square of the logarithm of the dimension of the input matrix. Therefore, plugging together the circuit of depth 1 which computes Step and the circuit computing transitive closure, and extracting as output the entry $\text{Step}^*_{\text{Init,Acc}}$ of the result, we obtain a circuit deciding $A \cap \{0,1\}^n$.

The depth of the circuit is $O(\log^2 N(n))$. Evaluating this expression, and using the fact that $S(n) \geq \log n$, we obtain that $d_A(n) \in O(S^2(n))$.   □

5.5 Polynomial Advice

In this section we use the boolean cost to characterize the class $P/poly$. The characterization uses the following definition:

Definition 5.22   *A set A has polynomial size circuits if and only if there is a polynomial p such that for each $n \geq 0$, $c_A(n) \leq p(n)$.*

It will be proved below that not every set has polynomial size circuits. The proof, as well as that of the characterization of $P/poly$, requires one first show that the evaluation of a circuit on a given input is not too difficult a task. In order to classify the problem of evaluating a circuit on an input, we define it formally:

Definition 5.23   *The circuit value problem,* CVP *for short, is the set of all pairs $\langle x, y \rangle$ where $x \in \{0,1\}^*$ and y encodes a circuit with $|x|$ input gates which outputs 1 on input x.*

Now we prove that the evaluation of a circuit on a given input can be done in time bounded by a polynomial in the size of the circuit. We will use this result in the characterization of $P/poly$.

Theorem 5.24 $CVP \in P$.

Proof.   During the evaluation of the circuit, we mark the evaluated gates. We say that a gate is *ready* if all its input gates either are inputs to the circuit or have already been evaluated. Evaluated gates are no longer ready. We make sure that inputs to ready gates are boolean constants, by substituting the result of each evaluation as soon as it is known. The algorithm of Figure 5.5 evaluates the circuit.

The correctness of this algorithm follows from the fact that if a gate eventually gets ready then it eventually gets evaluated, and that if gates at depth t eventually get evaluated then gates at depth $t + 1$ eventually get ready; inductively, it is easily shown that in cycleless, correct circuits every gate gets evaluated.

For the time bound of this algorithm, notice that substituting a value for a gate requires just one scanning of y, which takes linear time. Hence, for the first "for" loop quadratic time suffices. The second loop is performed at

---

```
input ⟨x, y⟩
check that y is a sequence of correctly encoded
    gates, and has |x| inputs; if not then reject
for i := 1 to |x| do
comment: feed the input into the circuit
        substitute the iᵗʰ bit of x for all the occurrences
            of the input gate xᵢ in y
loop
comment: actual gatewise evaluation
        scan y looking for a ready gate
        if found then
            evaluate it
            mark it "evaluated"
            substitute its value in every occurrence of this
                gate (as input to another) in the circuit
        else exit
end loop
if there are unevaluated gates
        then reject
comment: wrong circuit, it has a cycle!
else accept if and only if the value of the output gate is 1
end
```

Figure 5.5   Evaluation of a circuit

---

most as many times as there are gates in the circuit (surely less than $|y|$), and each loop needs linear time to find a ready gate, constant time to evaluate and mark it, and linear time to substitute the result. Thus the second loop can also be performed in quadratic time.   □

We use this fact to show the relationship between boolean circuit complexity and the advice classes, as given in the following result.

Theorem 5.25   *A set A has polynomial size circuits if and only if $A \in P/poly$.*

Proof.   Let A be a set with polynomial size circuits. We will show that A is in $P/poly$.

Let C_n be the encoding of a polynomial size circuit which recognizes $A \cap \Sigma^n$. Thus $|C_n| \leq p(n)$ for some polynomial p. Define the advice function as $h(n) = C_n$. Hence, $|h(n)|$ is polynomial in n. For any length n, and for any x of length n, we have that $x \in A$ if and only if C_n outputs 1 on input

x, if and only if $\langle x, h(|x|) \rangle \in$ CVP. As CVP is in P, this shows that A is in $P/poly$.

On the contrary, let A be in $P/poly$. Then by definition there exists a set B in P and an advice function h such that $x \in A$ if and only if $\langle x, h(|x|) \rangle \in B$.

Since B is in P, there exists a polynomial time deterministic Turing machine M such that $B = L(M)$. By Theorem 5.19, the boolean cost of B is polynomial. Therefore, for each length m there is a circuit which outputs 1 if and only if its input is in B.

As pairing functions are assumed to be polynomial time computable, by Corollary 5.20 they can be computed by polynomial size circuits. For each length n, construct a circuit with n inputs as follows: connect the n inputs and $|h(n)|$ constant gates corresponding to the actual value of $h(n)$ to the circuit computing the pairing function, to obtain $\langle x, h(n) \rangle$, and plug its output into the inputs of a circuit for B at length $m = |\langle x, h(n) \rangle|$. This circuit outputs 1 if and only if x is in A. As m is bounded by a polynomial of n, and all the circuits are of size polynomial in m, the whole circuit has size polynomial in n. □

This is a classical characterization of $P/poly$. Other characterizations of this class exist, and we now give one of them (others are proposed in Exercises 9 and 10). Our aim in the next theorem is to clarify the relationship between obtaining external information from polynomially long advices and obtaining it from oracles.

Theorem 5.26 $P/poly = \bigcup_{S \; sparse} P(S)$.

Proof. We prove that a set A is in $P/poly$ if and only if there is a sparse set S such that $A \in P(S)$. Assume that A is in $P/poly$ via the advice function f and the set B in P. Define the set S as follows:

$$S = \{ \langle 0^n, x \rangle \mid x \text{ is a prefix of } f(n) \}$$

We show that S is sparse. Each word in S of length m is of the form $\langle 0^n, x \rangle$ for $n \leq m$; there are $m + 1$ possible values of n. Each of them contributes at most $m + 1$ different prefixes of $f(n)$, one for each length up to m. The total number of words of length m is at most $O(m^2)$.

The set A is in $P(S)$, because the program in Figure 5.6 decides A in polynomial time using S as an oracle. Conversely, assume that $A \in P(S)$ where S is sparse. Let p be the polynomial bounding the running time of the machine that decides A. Define the advice function such that, for each n, it gives the encoding of the set of words in S up to size $p(n)$; this is a polynomially long encoding. Modify the machine so that it receives two inputs, x and $f(|x|)$, and then uses the set encoded in $f(|x|)$ for answering the queries to the oracle, instead of performing actual queries. This is a

```
input x
n := |x|
z := λ
loop
      if ⟨0ⁿ, z0⟩ ∈ S then z := z0
      else if ⟨0ⁿ, z1⟩ ∈ S then z := z1
      else exit
end loop
comment: z is now f(n)
if ⟨x, z⟩ ∈ B then accept
else reject
end
```

Figure 5.6   Prefix-searching construction of an advice

polynomial time machine which decides A with the polynomially bounded advice function f.   □

In Exercise 9 it is shown that the tally sets can substitute the sparse sets in this characterization.

We shall end this section by showing that not every recursive set has polynomial size circuits. This result will acquire its full importance in the light of Theorem 5.36, which states that "most" sets have exponential circuit size complexity.

Theorem 5.27   *There exist sets in EXPSPACE not having polynomial size circuits.*

Proof.   The proof consists of constructing a set A which agrees with the statement of the theorem. This construction will be performed in stages. Denote by x_i^n the ith string of length n, according to lexicographical order.

At stage n, the sets $\emptyset = A_{n,0} \subseteq A_{n,1} \subseteq \cdots \subseteq A_{n,2^n}$ are constructed. Each of these sets is equal either to the preceding set, or to the preceding set plus one string, so that $A_{n,i} - A_{n,i-1}$ is either $\emptyset$ or x_i^n. Finally, we define

$$A = \bigcup_{n \geq 0} A_{n,2^n}$$

The specific procedure for constructing $A_{n,2^n}$ is given in Figure 5.7.

We show that the construction produces the desired set. Observe that circuits of polynomial size have encodings of polynomial length. Fix any polynomial p. We show that it is not possible to decide correctly the set $A \cap \{0,1\}^n$ by circuits having encodings of length $p(n)$. Since p is arbitrary, this implies that A does not have polynomial size circuits.

stage n
$A_{n,0} = \emptyset$
for $i := 1$ to 2^n do
comment: to run over all the words of length n
    $total := 0$
comment: number of circuits considered so far
    $accep := 0$
comment: number of such circuits accepting x_i^n
    for each $w \in \{0,1\}^*$, $|w| \leq n^{\log n}$ do
        if w encodes a circuit c with n inputs which
        decides $A_{n,i-1}$ correctly on all the previous
        words $x_1^n, x_2^n, \ldots, x_{i-1}^n$ then
            $total := total + 1$
            if circuit c outputs 1 on x_i^n then $accep := accep + 1$
        end for w
    if $accep > total/2$
        then $A_{n,i} := A_{n,i-1}$
        else $A_{n,i} := A_{n,i-1} \cup \{x_i^n\}$
    end for i
end

Figure 5.7   Diagonalizing out of *P/poly*

Let n be the first natural number such that

$$p(n) < n^{\log n} < 2^n$$

and consider what happens at each loop of the outermost "for" loop at stage n. Since the first inequality holds, every potential encoding of a circuit for A having length $p(n)$ is considered in the innermost "for" loop. There are no more than $2^{n^{\log n}}$ encodings of circuits of length smaller than $n^{\log n}$. Initially all of them could be correct for A at length n; however, the (non)membership of x_1^n to A is determined contradicting the answer of at least half the correct circuits, so that fewer than half the circuits remain correct. Similarly, at each loop, the (non)membership of a word x_i^n to A is determined contradicting the answer of at least half the *remaining* correct circuits, so that again the number of correct circuits is halved. After this "spoiling" process has taken place through the full outermost loop, the initial upper bound of $2^{n^{\log n}}$ possible circuits has been halved 2^n times, hence divided by 2^{2^n}. By the second inequality, this number has been reduced to zero, and therefore no circuit of this size can decide $A_{n,2^n}$ correctly.

To prove that $A \in EXPSPACE$, we have to evaluate the complexity of recognizing the set A. In order to decide whether any x with $|x| = n$ belongs to A, it suffices to perform stage n of the construction. This stage consists of three nested exponentially bounded loops, where the innermost loop is the evaluation of circuit c on each of the inputs $x_1^n, x_2^n, \ldots x_{i-1}^n$. This can be done in time (hence space) polynomial in the length of the encoding of the circuit. Moreover, each iteration of the two inner loops can be performed reusing the same space as the previous ones. From an easy rough estimation it follows that A can be recognized in space $2^{O(n)}$ and time $2^{2^{O(n)}}$. □

Notice that the proof of this theorem shows a somewhat stronger result: it constructs a set such that for all but finitely many n, it cannot be decided at length n by circuits of size $n^{\log n}$. Polynomial size and infinitely many n would suffice for the statement.

Before going on to the next section, let us explain at this point the reason for the name "nonuniform complexity". Consider any way of describing finite sets by finite strings. There is a trade-off between encoding them in a very transparent and "readable" form, and encoding them in some highly compact way. For example, such an encoding could be a boolean circuit accepting exactly the strings in the set. The trade-off consists of the fact that in the first case a long encoding is obtained, but deciding whether a string is in the set encoded by another string is an easy searching job. On the contrary, in the second case the encoding string might be much smaller but an important amount of resources have to be spent on deciding whether a string is in the set encoded by another string.

Now fix a way of describing finite sets by strings, i.e. fix an algorithm (an "interpreter" of the descriptions) for deciding whether a string is in the set encoded by another. The complexity of a finite set is then the size of the smallest such description of the set. The "nonuniform complexity" of an infinite set A will then be the function which for each natural number n has as value the complexity of $A \cap \Sigma^n$. This is the intuitive idea behind the concepts studied in this chapter.

5.6 Logarithmic Advice

We turn now to the study of the classes defined by advice functions. Although the classes P/log and $P/poly$ are defined similarly, their properties are very different. Both are interesting classes which allow one to prove important results about uniform complexity classes. The class $P/poly$ has been characterized in the previous section, and we shall use it later on to develop several results. The class P/log also suggests important and useful ideas.

We begin with the consequences of the hypothetical fact $SAT \in P/log$.

Theorem 5.28 *If $SAT \in P/log$, then $P = NP$.*

Proof. The proof exploits the disjunctive self-reducibility structure of SAT. Let f be a logarithmic advice function, with $|f(n)| \leq c \cdot \log n$, and let B be a set in P, such that $F \in SAT$ if and only if $\langle F, f(|F|) \rangle \in B$.

Recall the notation $F|_{x:=c}$ for substituting the boolean constant c in all occurrences of x in F. Without loss of generality, we can assume that the resulting formula is padded out with some meaningless symbols in such a way that the length of $F|_{x:=c}$ is $|F|$.

Consider the program of Figure 5.8. It holds in variable z a formula which is always obtained by assigning values to variables in F. Thus, if this program accepts, then F is in SAT. On the other hand, if F is in SAT

```
input F
for each w with |w| ≤ c · log |F| do
    z := F
    while there is at least one variable in z do
        let x be the first variable in z
        if ⟨z|_{x:=0}, w⟩ ∈ B then z := z|_{x:=0}
        else if ⟨z|_{x:=1}, w⟩ ∈ B then z := z|_{x:=1}
        else exit the while loop
    end while
    if z simplifies to "true" then halt and accept
comment: a satisfying assignment has been found
end for
halt and reject
end
```

Figure 5.8 Deciding SAT with logarithmic advice

---

then for the value of $w = f(|F|)$ (which, by the way, is not recognized by the program) we have that B answers correctly, and correctly leads us to a satisfying assignment. The running time is easily seen to be polynomial in $|F|$.   □

We will prove in Volume II, that the hypothesis SAT $\in P/poly$ also has consequences for some uniform classes, although not as strong as shown for SAT $\in P/log$.

The proof technique we have seen is interesting, and will be used later: an advice taken from a polynomial amount of different advices can be useless, if there is a way of cycling through all of them and we can ensure that the "wrong" ones do not fool us. In this case, this was ensured by the self-reducibility structure of SAT.

We now present a characterization of the class P/log. Its interpretation could be that logarithmic advice does not provide actual nonuniformity, but just a "pointer" to some unknown object in a family of uniformly generated objects.

Theorem 5.29 *A set A is in the class $P/\log$ if and only if there is a set B in P, a polynomial time computable function f from $\{0\}^*$ to Σ^*, and a polynomial p, such that*

$$\forall n \exists i \leq p(n) \forall x, |x| = n, x \in A \text{ if and only if } \langle x, f(0^i) \rangle \in B$$

Proof. Let A be in $P/\log$, via the function h and the polynomial time decidable set B. Without loss of generality, we can assume that $h(n)$ has the form $h(n) = \langle h'(n), n \rangle$ for some h', where n is written in binary. Denote $B_m = \{x \mid \langle x, m \rangle \in B\}$. As B itself is in P, then for each fixed m, $B_m \in P$. Hence, by Theorem 5.19, for every n there exists a boolean circuit of size polynomial in n which decides B_m on inputs of length n. Define the function $f(0^m)$, where $m = \langle i, n \rangle$, as the encoding of the circuit that decides B_m on inputs of length n.

Fix n; then for each word x of length n, $x \in A$ if and only if $\langle x, h(|x|) \rangle \in B$, which is true if and only if $x \in B_{h(|x|)}$; hence, $x \in A$ if and only if the circuit $f(0^{h(|x|)})$ accepts x. Thus, for $i = h(n)$, $x \in A$ if and only if $\langle x, f(0^i) \rangle \in$ CVP, the circuit value problem, which is in P as shown in Theorem 5.24. Observe that $|i| \leq c \cdot \log_2 n$ implies that $i \leq p(n)$ for some polynomial p.

Conversely, assume that for each n there is an i less than or equal to $p(n)$, such that the knowledge of $f(0^i)$ allows B to decide A in the length n. Define $h(n)$ as the smallest such i. Then $|h(n)| = |i| \leq |p(n)| \leq c \cdot \log n$.

Define B' as follows: $B' = \{\langle x, i \rangle \mid i \leq c \cdot \log n \text{ and } \langle x, f(0^i) \rangle \in B\}$. Then, h and B' witness $A \in P/\log$, because $x \in A$ if and only if $\langle x, h(|x|) \rangle \in B'$.   □

Intuitively, this result says that $P/\log$ is the class of sets A for which we have a polynomial time computable collection of circuits, such that at least one among the first $p(n)$ of them computes A for the length n.

This result suggests a simple way of defining in similar terms a reducibility: we can use logarithmic advice also to compute reductions. The reducibility obtained plays a crucial role in our proof of an important result (Mahaney's Theorem) in Volume II.

Definition 5.30 *The set A is $(P/\log)$-m-reducible to the set B (which is denoted $A \leq_m^{P/\log} B$) if and only if there is a function f from $\{0\}^*$ into Σ^*, logarithmically bounded, and a polynomial time computable function g such that for all n, and for all x with $|x| \leq n$, $x \in A$ if and only if $g(\langle x, f(0^n) \rangle) \in B$.*

The main result for this reducibility, which we shall use later in Volume II, is that the reduction class of NP enjoys a property of closure under complements when restricted to sparse sets. This result is a useful technical tool.

Theorem 5.31   *Let S be sparse, and assume that $S \leq_m^{P/log}$ SAT. Then $\overline{S} \leq_m^{P/log}$ SAT.*

Proof. Let f, g be functions as in the definition for the reducibility from S to SAT; assume the length of $f(0^n)$ to be bounded by $k \cdot \log n$. Consider the nondeterministic program of Figure 5.9.

```
input ⟨x, ⟨w, c⟩⟩
check that |w| ≤ k·log|x| and that c ≤ p(|x|)
nondeterministically guess c words, x₁,...,x_c,
      each of them of length at most |x|
check that the guessed words are all different
check that x is different from all the guessed words
for each i do
      nondeterministically check that g(⟨x_i, w⟩) ∈ SAT
if all the conditions hold then accept
end
```

Figure 5.9   A nondeterministic algorithm for a co-sparse set

---

It is easy to see that this program runs in time polynomial in $|x|$, and hence the language it accepts belongs to NP. Let g' be a polynomial time reduction from this language to SAT. Observe that if it is given the value of $f(0^{|x|})$ as input w, and if it is given the value of the census $C_S(n)$ as input c, then the program accepts if and only if the input x is *not* in S. Thus, defining $f'(0^n) = \langle f(0^n), C_S(n)\rangle$, we have that $x \in \overline{S}$ if and only if $\langle x, f'(0^{|x|})\rangle$ is accepted by the program, and as g' is a reduction to SAT, this happens if and only if $g'(\langle x, f'(0^{|x|})\rangle) \in$ SAT. Therefore, $\overline{S}$ is P/log reducible to SAT, which was to be shown. $\square$

5.7 Self-Producible Circuits

It is reasonable to consider sets with polynomial size circuits, in which some condition is imposed on the possibilities of computing these circuits. Some of these cases are proposed in the Exercises. We present here an interesting case, which can be characterized in an elegant way, and which will be used in Volume II.

From the results in the previous sections, circuits can be identified with a polynomially long advice string. Let A be a set in $P/poly$, via some advice function f. This means that f encodes information about A which can be easily recovered (i.e., recovered in polynomial time). This suggests the following questions: what is the relationship between this advice function and the set A considered as an oracle? Is it possible to compute easily f using an oracle for A? We shall characterize the sets for which this is possible. First, we define the concept in a more formal way.

Definition 5.32 *A set A has self-producible circuits if and only if there is a set B in P and a function f from $\{0\}^*$ to Σ^*, which is computable in polynomial time using oracle A, such that $x \in A$ if and only if $\langle x, f(0^{|x|}) \rangle \in B$.*

Thus, a set A has self-producible circuits if and only if it is in the class $P/PF(A)$. Observe that this class depends on the set A. Some classes of sets have been shown to have this property. In other parts of this book, we show additional facts about them, and use these facts. For the moment, we characterize them by an interesting property: they are precisely the sets belonging to a tally polynomial T-degree. (Following a standard convention, we say that a degree is a *tally polynomial degree* if and only if it is the polynomial degree of a tally set.)

Theorem 5.33 *A set A has self-producible circuits if and only if there is a tally set T such that $A \equiv_T T$.*

Proof. Let A have self-producible circuits, via function f and set $B \in P$. Let p be a polynomial bounding the running time of the machine computing f with oracle A, and hence bounding the length of f. Form the following tally set T:

$$T = \{0^n \mid n = \langle i, j, 3 \rangle, \text{ and } f(0^i) \text{ has at least } j \text{ bits } \} \bigcup$$

$$\{0^n \mid n = \langle i, j, 4 \rangle, f(0^i) \text{ has at least } j \text{ bits with the } j^{\text{th}} \text{ bit a } 0\}$$

First we claim that $A \in P(T)$. To show this, define a machine that works as indicated in Figure 5.10. By the definition of self-producible circuits, this machine accepts exactly A.

Now we claim that $T \in P(A)$. To show this, define a machine that works as indicated in Figure 5.11. By the definition of T, this proves the forward direction.

For the converse, assume that A and T are equivalent, where T is any tally set. Let M_1, M_2 be polynomial time oracle machines such that $A = L(M_1, T)$ and $T = L(M_2, A)$; let p_1 and p_2 be the polynomials bounding their respective running times. The algorithm of Figure 5.12 "constructs circuits" for A (i.e., evaluates an advice function f for A) in polynomial time with the aid of A itself as oracle. The construction is easily seen to work in polynomial time with oracle A. For this advice function, the appropriate set $B \in P$ is defined by the machine that computes the circuit value problem, i.e. evaluates a given circuit on a given input. Since it has been shown in Theorem 5.24 that CVP is in P, this completes the proof.   □

```
input x
z := λ
for each j between 0 and p(|x|) do
      let n = ⟨|x|, j, 3⟩
      query T about 0ⁿ
      if the answer is YES then
            let m = ⟨|x|, j, 4⟩
            query about 0ᵐ
            if the answer is YES then z := z0 else z := z1
comment: z is now f(0^|x|)
check if ⟨x, z⟩ ∈ B and answer accordingly
end
```

Figure 5.10   A Turing reduction to a tally set

Corollary 5.34   *If A has self-producible circuits and $DEXT(A) = DEXT$, then $A \in P$.*

Proof.   From Exercise 22 in Chapter 4, we know that for tally sets T, if it holds that $DEXT(T) = DEXT$ then $T \in P$. Thus, if $DEXT(A) = DEXT$ and $P(A) = P(T)$, then $DEXT(T) = DEXT(A) = DEXT$; hence $T \in P$, and therefore $A \in P$.   □

A particular subclass of the sets with self-producible circuits is studied in Volume II of this book.

```
input x
if x is of the form 0ⁿ with n = ⟨i, j, 3⟩ then
      compute with the aid of oracle A the value of f(0ⁱ)
      and accept if it has at least j bits
else if x is of the form 0ⁿ with n = ⟨i, j, 4⟩ then
      compute with the aid of oracle A the value of f(0ⁱ)
      and accept if it has at least j bits and the
            jᵗʰ bit is a 0
else reject
end
```

Figure 5.11   A Turing reduction from a tally set

input 0^n
for each j between 0 and $p_1(n)$ do
    simulate M_2 on each input 0^j (using the oracle
        for solving the queries)
    in this way, construct a circuit accepting exactly
        T up to length $p_1(n)$
construct and output a circuit simulating M_1 on inputs of
    length n, plugging in repeatedly the circuit for T in
    order to compute the next configuration when M_1 enters
    a query configuration
end

Figure 5.12   Producing circuits for a set

5.8 A Lower Bound to the Circuit Size of Boolean Functions

In this section we shall prove that "most" of the boolean functions have a circuit complexity near $2^n/n$. The proof of the result uses a purely combinatorial argument, and does not give us information about the structure of the boolean functions which admit such a lower bound. In some sense, this result complements Theorem 5.27, where we constructed a recursive set which does not have polynomial size circuits.

Let us start with some definitions. For any fixed function g from $\mathbb{N}$ to $\mathbb{N}$, denote by $H_g(n)$ the number of functions $f : \{0,1\}^n \to \{0,1\}$ such that $c(f) \leq g(n)$.

Since the number of boolean functions on n variables is 2^{2^n}, the fraction of functions with cost bounded by $g(n)$, denoted $F_g(n)$, will be

$$F_g(n) = \frac{1}{2^{2^n}} H_g(n)$$

Let us start deriving an upper bound for $H_g(n)$. To do this, we shall use the following weak version of the well-known Stirling formula: $n! \geq n^n \cdot e^{-n}$.

Lemma 5.35   *If $g(n) \geq (n+1)$ then $H_g(n) \leq (12 \cdot e \cdot g(n))^{g(n)}$*

Proof.   Let f be any boolean function on n arguments. Whenever $c(f) \leq g(n)$, there exists a circuit of size exactly $g(n)$ synthesizing the function f. This circuit has the structure shown in Figure 5.13, where C_1 computes f with a cost $c(f)$, and C_2 is a useless padding circuit of size $g(n) - c(f)$.

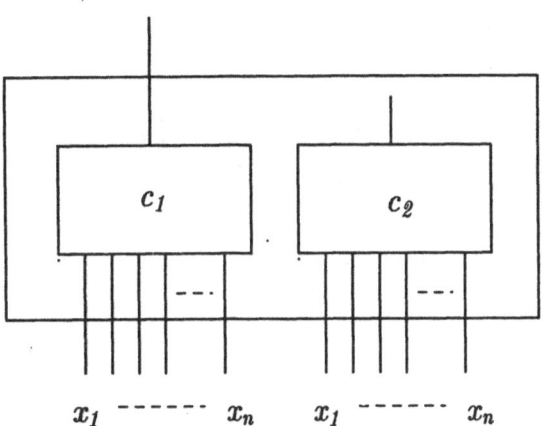

Figure 5.13   Padding a boolean circuit

After this remark, it should be clear that to bound $H_g(n)$ it is sufficient to study the number of boolean functions that can be synthesized with exactly $g(n)$ gates. To count such a number, we begin by considering the number $G_g(n)$ of different unlabeled graphs corresponding to circuits with n inputs and $g(n)$ internal nodes. To get an upper bound on $G_g(n)$, consider the set of internal nodes $\{1, 2, 3, \ldots, g(n)\}$, and the set of input nodes $\{x_1, x_2, \ldots, x_n, 0, 1\}$. Each internal node i is connected to at most two other nodes j and k. We can choose node j out of $g(n) + n + 1$ possibilities, and the same for node k. Therefore there are $(g(n) + n + 1)^2$ possibilities to assign the neighbors of i. But by hypothesis $g(n) \geq (n + 1)$, so that $(g(n) + n + 1)^2 \leq (2 \cdot g(n))^2$. As the total number of internal nodes is $g(n)$, an upper bound to $G_g(n)$ is $(2 \cdot g(n))^{2 \cdot g(n)}$. However, this way of counting gives us the same graph $g(n)!$ times, therefore we can sharpen the estimation to obtain

$$G_g(n) \leq \frac{1}{g(n)!} \cdot (2 \cdot g(n))^{2 \cdot g(n)}$$

We have found an upper bound to the total number of unlabeled graphs which corresponds to a circuit with n inputs and $g(n)$ internal gates. To transform these graphs in actual circuits, we need to label each of the internal nodes with a boolean operation from $\{\wedge, \vee, \neg\}$. As there are $g(n)$ internal nodes, we obtain

$$H_g(n) \leq 3^{g(n)} \cdot G_g(n) \leq \frac{1}{g(n)!} 12^{g(n)} g(n)^{2g(n)}$$

and by use of the weak Stirling formula we get $H_g(n) \leq (12 \cdot e \cdot g(n))^{g(n)}$.   □

Now we can proceed to the main theorem of this section. It says that the ratio of functions having a boolean cost bounded below (and kept constantly away from) $\frac{2^n}{n}$ is as small as desired for large enough n.

Theorem 5.36   *For every* ε , $0 < \varepsilon < 1$, *if* $g(n) = (1 - \varepsilon) \cdot \frac{2^n}{n}$, *then*

$$\lim_{n \to \infty} F_g(n) = 0$$

Proof.   Substituting $\dot{g}$ in Lemma 5.35, we obtain by straightforward manipulation that

$$H_g(n) \le \left(\left(\frac{1}{n} \right) (12 \cdot e \cdot (1 - \varepsilon)) \right)^{g(n)} \cdot 2^{(1-\varepsilon) \cdot 2^n}$$

As $F_g(n) = H_g(n)/(2^{2^n})$, we obtain

$$F_g(n) \le \left(\left(\frac{1}{n} \right) (12 \cdot e \cdot (1 - \varepsilon)) \right)^{g(n)} \cdot 2^{(-\varepsilon) \cdot 2^n}$$

and since $\lim_{n \to \infty} g(n) = \infty$, we get that $\lim_{n \to \infty} F_g(n) = 0$.   □

5.9 Other Nonuniform Complexity Measures

We review in this section some work on other nonuniform complexity measures, just to give the reader a feeling for the kind of results that can be obtained in this approach.

We define first some recently used measures. We will show only one result about one of them, but all of them are quite interesting for the complexity-theorist.

Definition 5.37   *The fan-out of a gate g of a boolean circuit is the number of gates having the output of g as input. The fan-out of a boolean circuit is the maximum fan-out of its internal (non-input) gates.*

It can be proved that bounding the fan-out of circuits by a constant greater than 1 does not increase the size of the circuit by more than a polynomial. However, it is not known whether circuits of unbounded fan-out are more powerful than circuits of fan-out 1.

Boolean formulas were defined in Chapter 1. The following relationship holds:

Proposition 5.38   *A subset A of $\{0,1\}^n$ is accepted by a circuit with k internal gates and fan-out 1 if and only if A is the set of satisfying assignments of a boolean formula with k connectives.*

Proof. Given a circuit with k internal gates and fan-out 1, observe that the internal gates have a tree-like structure. Construct a formula as follows: consider the output gate, observe the boolean operation performed at the gate, and use the corresponding connective to connect at most two formulas constructed recursively from the right and left subtrees, which are again circuits. Input gates correspond to literals in the formula. Conversely, given a formula, write it as a tree in which each node corresponds to a connective, and has as sons the trees corresponding to the formulas it connects. This tree gives the structure of the boolean circuit with fan-out 1 equivalent to the formula.   □

In a similar manner to the definition of a set having polynomial size circuits, the concept of *polynomial size formulas* can be defined. Of course, Proposition 5.38 implies that if a set has polynomial size formulas then it has polynomial size circuits. The converse is not known to hold. Other boolean computation models exist (quantified boolean formulas, branching programs, decision trees, etc.), but their study in depth falls outside the scope of this book. See the References for characterizations of the classes defined by some of these computation models.

Automata theoretic and language theoretic tools can also be used to define nonuniform complexity measures. We present the most basic of them; others can be found in the references mentioned in the Bibliographical Remarks.

Definition 5.39   *Given a subset A of $\{0,1\}^*$, consider for each n the smallest nondeterministic finite automaton M_n accepting $A \cap \{0,1\}^n$. Denote by $a_A(n)$ the number of states of M_n. The function $a_A(n)$ is called the initial index of A. A set has polynomial initial index if and only if $a_A(n) \leq p(n)$ for some polynomial p.*

It can be shown by counting arguments that some sets do not have polynomial initial index. See Exercise 22. The class of sets having polynomial initial index can be characterized in a similar manner to the class of sets with polynomial size circuits.

Recall the definition of on-line Turing machines from Chapter 1. A logspace on-line deterministic or nondeterministic machine is started in an initial configuration in which the input x is written on the input tape, its length $|x|$ is written in binary on a work tape, and $\log |x|$ tape cells are laid off in the remaining work tapes. The machine is not allowed to use more space than marked on the tapes.

These machines may also query an oracle; in this case the space for the query tape is unbounded.

For an oracle A, the class $NLOG_{on}(A)$ is the class of sets decided with the help of oracle A by nondeterministic log-space on-line Turing machines, and $DLOG_{on}(A)$ is the corresponding deterministic class. These classes allow one to characterize the sets with a polynomial initial index.

Theorem 5.40   *A set A has polynomial initial index if and only if A belongs to $NLOG_{on}(S)$ for some sparse set S.*

Proof.   Let A have polynomial initial index; then there is a sequence of nondeterministic finite automata M_n accepting $A \cap \{0,1\}^n$. Denote by S the set of 5-tuples $\langle 0^n, x, i, j, u \rangle$ such that i and j are states of M_n, x is an input symbol, M_n can transfer from state i to state j when reading symbol x, and u is 1 if j is a final state of M_n and 0 otherwise. The fact that S is sparse is proven exactly as in Theorem 5.26. Using this oracle, A can be accepted in nondeterministic on-line logarithmic space as indicated in Figure 5.14.

$i :=$ initial state of M_n
comment: i denotes the current state
loop
      read next symbol x of the input
      guess j with $|j| = |i|$
comment: j denotes the next state
      guess whether j is a final state, i.e. $u := 0$ or $u := 1$
      query S about $\langle 0^n, x, i, j, u \rangle$
      if the answer is YES then $i := j$
      else reject
comment: wrong guess
until the whole input has been read
accept if and only if $u = 1$
end

Figure 5.14   A nondeterministic log-space on-line machine

The value of n is the length of the input, which is known by the machine. It is immediately clear that this procedure accepts A on-line in nondeterministic logarithmic space with oracle S.

Conversely, let A belong to $NLOG_{on}(S)$, and let M be a machine witnessing this fact. Observe that M runs in polynomial time, and hence queries about words of polynomial length. Moreover, the number of work tape configurations of M is polynomial. Let S_n be a sequence of finite polynomial size automata accepting the words of S that can be queried by M on inputs of length n. We construct automata M_n whose states are triples (q, C, r), where q is a state of M, C is a possible work tape configuration of M on an input of length n, and r is a state of S_n. The transitions of M_n follow the transitions of M when M does not query or write on the query tape. When q is the query state, the transition leaves C unchanged, sets q to the YES answer state if r was a final state of S_n and to the NO answer state otherwise, and finally sets r to the initial state of S_n. When M writes symbol x on the query tape, q and C are updated according to the transition function of M and r is updated according to the transition function of S_n.

We obtain a polynomial size nondeterministic finite automaton accepting $A \cap \{0,1\}^n$. This automaton has λ-transitions, which can be eliminated by standard means with a polynomial overhead on the number of states.   □

Interesting corollaries follow from this characterization (see Exercise 23). Similar results can be proved for deterministic finite automata (see Exercise 21) and for other models like pushdown automata. Context-free and context-sensitive grammars have also been used to measure nonuniform complexity. For many nonuniform measures, there exist characterizations in terms of complexity classes relative to sparse oracles. Throughout this book, however, boolean circuits will be much more important than any other nonuniform measure.

5.10 Exercises

1. Consider the class PF of functions which are computable within polynomial time. Show that $P/PF = P$.
2. Show that $(P/poly)/poly = P/poly$. Look for sufficient conditions on C and $\mathcal{F}$ for $(C/\mathcal{F})/\mathcal{F} = C/\mathcal{F}$.
3. Show that $(\text{co-}C)/\mathcal{F} = \text{co-}(C/\mathcal{F})$.
4. Prove Corollary 5.20.
5. Show that no T-hard set for $EXPSPACE$ can have polynomial size circuits.
6. What is the closure of P/log under polynomial time Turing reducibility? Define a variant of P/log for which you can prove that it is closed under polynomial time Turing reducibility. Characterize it in a manner similar to that of Theorem 5.29.
7. Compare the class defined in the previous exercise with the minimum nontrivial degree of the (P/log)-m-reducibility.
8. Show that the (P/log)-m-reducibility is a preorder (i.e. is a reflexive and transitive relation).
9. Show that a set has polynomial size circuits if and only if it is in $P(T)$ for some tally set T. *Hint:* follow the direct implication of the characterization of sets with self-producible circuits.
10. Show that a set has polynomial size circuits if and only if it is truth-table reducible in polynomial time to some tally set.
11. A set A is called *p-selective* if there is a function f, called the *selector function*, which is computable in polynomial time and such that for all x,y in Σ^* we have the following properties:

    (a) $f(x,y) = x$ or $f(x,y) = y$,
    (b) if x is in A or y is in A then $f(x,y)$ is in A.

Prove that if A is a p-selective set with selector function f, then for each $n \geq 0$, there exists a set C in $A \cap \Sigma^n$ with $|C| \leq n + 1$ such that

$$A \cap \Sigma^n = \{x \mid |x| = n \text{ and } \exists y \in C \text{ such that } f(x,y) = x \text{ or } f(y,x) = x\}$$

12. Using Exercise 11, prove that each p-selective set has polynomial size circuits.

13. Let A be a p-selective set. Prove that $\overline{A}$ is also p-selective and that $A \times A \leq_m A$.

14. Prove that for every set A, $A \in P$ if and only if A is p-selective and $A \leq_{pos} \overline{A}$.

15. Prove that the circuit value problem is complete for P with respect to logarithmic space reducibility. *Hint:* Define f so that it performs the construction presented in Theorem 5.19.

16. •Polynomial time truth-table reducibility was defined in Exercise 10 of Chapter 4. Show that it can be defined equivalently as follows: $A \leq_{tt} B$ if and only if there are two functions f and g, both computable in polynomial time, such that, for each x, $f(x)$ is a list of words $w_1, w_2, \ldots, w_k$ and $x \in A \Leftrightarrow \langle \chi_B(w_1) \cdots \chi_B(w_k), g(x) \rangle \in \text{CVP}$. Here $\chi_B(w_1) \cdots \chi_B(w_k)$ represents the binary word obtained by ordered concatenation of all the bits $\chi_B(w_i)$.

17. •Show that $DEXT \neq EXPSPACE$ if and only if $PSPACE \cap P/poly \neq P$. *Hint:* Combine Theorem 5.26 with the ideas leading to the proof of Corollary 4.22.

18. A *generator* is a circuit G with m input gates and n output gates, computing a boolean function f from $\{0,1\}^m$ to $\{0,1\}^n$, which has an extra output gate called "domain indicator". The output of G is considered undefined whenever the domain indicator outputs 0. In this way, G computes a partial function which coincides with f when the domain indicator is 1, and is undefined otherwise. We say that G *generates* the set A if and only if A is the range of this partial function. A subset A of $\{0,1\}^*$ has *polynomial size generators* if for some polynomial p and each integer $n > 0$ there exists a circuit C_n with m inputs and at most $p(n)$ gates, such that C_n generates $A \cap \{0,1\}^n$

   (a) Show that if a set A has polynomial size circuits, then A has polynomial size generators.

   (b) Consider the following problem: given a generator G and a word w, decide whether w belongs to the range of G. Formalize this problem and prove that it is *NP*-complete.

19. •Show that the following are equivalent:

   (a) A has polynomial size generators.
   (b) $A \in NP/poly$.
   (c) There exists a tally set T such that $A \in NP(T)$.

(d)   There exists a sparse set S such that $A \in NP(S)$.

Hint: In (d) $\Rightarrow$ (a), use Theorem 5.19 to show that a polynomial size circuit can check that the input to the generator is the encoding of an accepting computation.

20.   Show that there exist sets in *EXPSPACE* not having polynomial size generators. *Hint:* Use a construction like that in the proof of Theorem 5.27.

21.   Define the *deterministic initial index* using deterministic finite automata. Show that a set A has polynomial deterministic initial index if and only if A belongs to $DLOG_{on}(S)$ for some sparse set S. *Hint:* In the proof of Theorem 5.40, substitute a prefix search for the guess of the next state, as in the proof of Theorem 5.26.

22.   Show that the set of palindromes with a central separator,

$$\left\{ w\#w^r \mid w \in \{0,1\}^* \right\}$$

does not have polynomial initial index.

23.   Use Exercise 22 to show that $NLOG_{on}$ is not closed under complements. Conclude that $DLOG_{on}$ is strictly included in $NLOG_{on}$.

5.11 Bibliographical Remarks

The study of the combinational complexity, or boolean circuit complexity, of functions dates back to the doctoral dissertation of C. Shannon (1938), in which he used boolean algebra for the design and analysis of switching circuits. In other papers he continued the development of boolean functions applied to the study of switching circuits—Shannon (1949). Due to this work, a group of Russian mathematicians, led by O. Lupanov, carried on a very successful program of research on the circuit complexity of boolean functions. Among many results, a particularly important one, known as Lupanov's Theorem—Lupanov (1958)— states that for almost all boolean functions f,

$$c(f) \leq \left(1 + O\left(\frac{1}{\sqrt{n}}\right)\right) \cdot \left(\frac{2^n}{n}\right)$$

—see also Theorem 3.4.3 of Savage (1976). This upper bound is tight, as shown by the lower bound on the cost of boolean functions presented in Theorem 5.36 in the text. Theorem 5.36 is the counterpart of Lupanov's Theorem, and is due to Riordan and Shannon (1942). Our proof is an adaptation of a proof presented in Savage (1976).

Since the early 70's, circuit complexity has been widely used to find lower bounds to the complexity of problems; see for example Harper and Savage (1972) and Borodin (1973). However, in spite of all the efforts in trying to prove an exponential lower bound for the circuit size of a problem in

NP (which would imply $P \neq NP$), the best lower bound known for such a problem is in the range of $3n$, found by Blum (1984). The book of Savage (1976) contains a detailed exposition of the most important results up to 1976.

It also should be mentioned that Savage's book gives a more general definition of circuit than the one presented in this book, in the sense that he defines circuit over any complete base, while we do it here on the specific base $\{\wedge, \vee, \neg\}$.

The relationship between Turing machine time and circuit size, which is presented in Theorem 5.19, is basically due to Savage (1972). Improvements of this trade-off can be found in Pippenger and Fischer (1979) and in Schnorr (1976b). Theorem 5.21 is due to Borodin (1977). Theorem 5.24 and Exercise 15 are from Ladner (1975b).

Our general definition of nonuniform complexity classes is taken from Karp and Lipton (1980). Similar concepts were previously defined by Plaisted (1977) and Pippenger (1979). Theorem 5.25 is presented in Pippenger (1979). Theorem 5.26 is an example of good oral transmission, and in the large number of papers where it has been used is attributed to A. Meyer. Theorem 5.27 is presented following Schöning (1985a), where he uses an argument from Kannan (1982); he also presents Exercise 20.

The results developed about the class *P/log* come from different sources. Theorem 5.28 is due to Karp and Lipton (1980), slightly reformulated. The characterization in Theorem 5.29, as well as the definition of *P/log-m*-reducibility, are, to our knowledge, new. However, Theorem 5.31, which is stated as a property of this reducibility, is just an argument from Mahaney (1982); actually it is a "slice" of the proof of its main theorem, formulated in terms of the *P/log-m*-reducibility.

The notion of self-producible circuits is due to Ko (1985). Theorem 5.33 is from Balcázar and Book (1986). Initial index was introduced by Gabarró (1983a and 1983b). Exercise 22 is proven there. The results about nonuniform measures presented in the last section, as well as Exercises 21 and 23, follow Balcázar, Díaz, and Gabarró (1985). In this reference, more information can be found about other measures. Similar characterizations of different nonuniform measures can be found in Balcázar, Díaz, and Gabarró (1987) and in Balcázar and Gabarró (1989).

The definition of *p*-selective set given in Exercise 11 is due to Selman (1979), and is a polynomial time analogue of the concept of "semirecursive" set introduced in Jockusch (1968). The *p*-selective sets were further used to distinguish between different kinds of polynomial time reducibilities on *NP*, giving rise to some fine results about the structure of the class *NP*. Some results of this type are Exercises 13 and 14, which are taken from Selman (1982a). Further results can be found in Selman (1982b). The results stated in Exercises 11 and 12 are from Ko (1983).

Exercise 17 is from Hartmanis and Yesha (1984). The concept of generator (Exercise 18) is due to Yap (1983). Exercise 19 is from Yap (1983) and Schöning (1984b).

6   Probabilistic Algorithms

6.1 Introduction

In recent times, probabilistic methods have become increasingly used in the design and analysis of algorithms. On the one hand, they have been applied to the analysis of the average case complexity of deterministic exact or approximate algorithms. On the other hand, (pseudo-)random number generators can be incorporated into an algorithm, in such a way that the algorithm will answer correctly with a reasonably high probability. These are called probabilistic, or randomized, algorithms.

Thus, a probabilistic algorithm is a procedure that behaves in a deterministic way, except that occasionally it takes decisions with a fixed probability distribution.

For a flavor of this kind of algorithm, we will give the outline of Rabin's probabilistic algorithm to test in polynomial time whether a given integer is a prime. This is one of the first known examples of probabilistic algorithms, and is related to the nondeterministic algorithm for primality given in Theorem 3.4.

Let n be a given natural number. For each integer x with $1 \leq x < n$, denote by $W(x)$ the following condition: either $x^{n-1} \not\equiv 1 \pmod{n}$, or there exists an integer m of the form $m = (n-1)/2^i$ such that $x^m - 1$ and n have a common divisor different from 1 and n.

The algorithm is based upon the following property: if $W(x)$ holds for some x, then n must be composite; and if n is composite, at least half of the integers x between 1 and n satisfy $W(x)$.

Now the algorithm is easily stated; see Figure 6.1. From the property stated above, if the algorithm accepts then n must be composite, because an x has been found for which $W(x)$ holds. On the other hand, if the algorithm rejects then n is "probably" a prime: n is prime only if *all* of the chosen x_j fail to pick a witness to the compositeness of n. As half the numbers are such witnesses, the probability of making such a mistake is at most $\left(\frac{1}{2}\right)^m$.

In this chapter we shall propose a formal computational model for probabilistic algorithms: the Probabilistic Turing Machine. We will use probabilistic machines to define and study complexity classes by imposing bounds

---

input n
choose randomly m integers, $x_1, x_2, \ldots, x_m$,
    such that $1 \leq x_j < n$
for each x_j, test whether $W(x_j)$ holds
if $W(x_j)$ holds for some x_j then accept
else reject
end

Figure 6.1 Rabin's probabilistic (non-)primality test

---

on the time they are allowed to spend on the computation. Also, the intuitive concept of the probability of error used in the discussion above will be formalized using this model.

6.2 The Probabilistic Computational Model

Our probabilistic model of computation is similar to the nondeterministic model. The difference is that instead of "guessing" the next move, we "toss-a-coin" and make the move as a function of the outcome. This makes a difference in the definition of accepting: while in nondeterminism the input is accepted if and only if there is at least one computation which finishes in an accepting state, in the probabilistic machines we consider the probability of getting an accepting computation. With each given computation we associate a probability which is the product of the probabilities at the coin-tossing steps of the computation. The probability of accepting an input is the sum of the probabilities associated with the accepting computations of the machine on the given input.

Hence, for our purposes, a *probabilistic Turing machine* is just a kind of nondeterministic Turing machine in which acceptance is defined in a different way. We impose the following conditions on the nondeterministic machines:

1. Every step of the computation can be made in exactly two possible ways, which are considered different *even if there is no difference in the corresponding actions*. Hence, in the course of the computation, every configuration of the machine has exactly two configurations allowed as its next configuration. This can be achieved by writing down on a separate work tape a 0 or a 1 at each step of the computation. If the step was deterministic, then this is the only difference between the two successor configurations.

2. The machine is "clocked" by some time constructible function, as described in Section 2.4, and the number of steps in each computation is *exactly* the number of steps allowed by the clock. If a final state is

reached before this number of steps, then the computation is continued, doing nothing up to this number of steps.

3. Every computation ends in a final state, which can be either ACCEPT or REJECT.

The function that clocks the machine is called its "running time". Observe that the last two conditions ensure that the tree of computations is a full binary tree having the value of the running time as height. A nondeterministic machine fulfilling the conditions above can be considered a probabilistic machine by redefining acceptance as follows: a word x is considered accepted by a probabilistic machine M if and only if *more than half* the computations of M on x end in the ACCEPT final state. The fact that all computations have the same length allows us to define acceptance equivalently as "the probability of finding an accepting computation is greater than $\frac{1}{2}$", because all computations have the same probability. Still another way of stating this fact would be "the ratio of accepting computations to the total number of computations is greater than $\frac{1}{2}$".

Note that "greater than" is used in the definition, and not "greater than or equal to". Hence, if exactly half the computations accept and therefore exactly half the computations reject, then the input word is considered rejected. However, it can be shown (Exercise 1) that neither this condition nor the value $\frac{1}{2}$ is crucial. Under certain conditions, they can be changed without changing the power of the polynomial time version of this model.

An important aspect of the probabilistic procedures is the reliability. We study this by defining the "error probability" of the machines, i.e. the probability that the machines answer incorrectly. In the next section, we will bound this probability to define classes of more reliable probabilistic algorithms.

Definition 6.1 *The error probability of a probabilistic machine M is the real-valued function $e_M(x)$ defined by the ratio of computations on input x giving the wrong answer, to the total number of computations on x.*

Thus, the error probability is the ratio of accepting computations on probabilistically rejected words, and the ratio of rejecting computations on probabilistically accepted words; both to the total number of computations. It is clear from the definition that $e_M(x)$ is always less than $\frac{1}{2}$.

Let us see an example of probabilistic computation.

Example 6.2 Let MAJ be defined as follows: it is the set of boolean formulas which are satisfied by more than half of the possible assignments. This set can be accepted by a probabilistic machine described in the following. Consider the nondeterministic algorithm for SAT given in Example 1.46 (see Figure 6.2). It is easy to transform this machine into a nondeterministic machine fulfilling the conditions imposed above so that it can be considered a probabilistic machine.

input F
check that F encodes a correct boolean formula
for each variable x occurring in F do
    choose with equal probability
        $F := F|_{x:=0}$ or
        $F := F|_{x:=1}$.
simplify the resulting formula without variables
if the formula simplifies to "true" then accept and halt
end

Figure 6.2 A probabilistic algorithm for MAJ

This machine accepts exactly those formulas for which more than half the computations are accepting. But each accepting computation uniquely corresponds to a satisfying assignment; thus, a formula is probabilistically accepted if and only if it is satisfied by more than half the possible assignments. Therefore this probabilistic machine accepts MAJ.

Now we turn to simulating a probabilistic machine by a deterministic one. Of course the fact that the full computation tree must be constructed implies that the running time "blows up" in the simulation.

Theorem 6.3 *Let t be a space constructible function. A probabilistic machine M with running time t can be simulated by a deterministic machine in space $O(t)$ and time $O(2^t t^2)$.*

Proof. Consider a deterministic machine M' which works as indicated in Figure 6.3.

Note that the for loop makes M' simulate every possible computation of M. Thus, at the end of the loop the counter indicates the number of accepting computations of M, and hence M' accepts x if and only if M accepts probabilistically x.

The space required by M' is the same space required by M (which is bounded by its running time t), plus t additional bits to hold w, plus t additional bits to keep the counter, which never exceeds 2^t, and can be written in binary with t bits. Thus the total space is $O(t)$.

For a space constructible t, the running time of a machine which uses space t can be bounded by $2^{O(t)}$, as indicated in the proof of Lemma 2.25. Closer inspection gives the stated time bound, which can be refined further.
□

```
input x
initialize a counter to 0
for each word w over {0,1} of length t(|x|) do
      simulate step by step a computation of M on x, reading
            a bit of w at each step, by walking through the
            computation tree of M, going left if the bit of w is 0,
            and right if it is 1
      if the computation accepts then increment the counter by 1
end for
if the counter is greater than 2^(t(|x|)-1) then accept
      else reject
end
```

Figure 6.3   Deterministic simulation of a probabilistic machine

---

6.3 Polynomial Time Probabilistic Classes

We now start to study the complexity classes defined by probabilistic Turing machines. We shall consider polynomial time bounds, and different bounds on the error probability. A polynomial time bounded probabilistic machine is a probabilistic machine whose running time is a polynomial $p(n)$. This implies that every computation of the machine on inputs of length n halts in exactly $p(n)$ steps.

Definition 6.4   *We denote by PP the class of languages accepted by polynomially clocked probabilistic Turing machines.*

The name PP stands for **P**robabilistic **P**olynomial time. Let us locate the class PP in our realm of interest.

Proposition 6.5 $NP \subseteq PP \subseteq PSPACE$.

Proof.   By the simulation in Theorem 6.3, polynomial time probabilistic machines can be simulated in polynomial space. Hence $PP \subseteq PSPACE$.

To prove the other inclusion, let $L \in NP$. Then there exists a nondeterministic machine M with $L = L(M)$. Let M' be the nondeterministic machine of Figure 6.4.

It is an easy task to transform M' into a machine M'' fulfilling the conditions in the definition of probabilistic machines, by extending the short computations with "do-nothing" steps, and splitting into two identical configurations at each step. Let us prove that the set accepted probabilistically by M'' is L.

If $x \notin L$, then exactly half the computations accept x: those which start by choosing option 1 in the algorithm. Computations starting with option

input x
choose nondeterministically to perform either 1 or 2
    1: accept without further computation
    2: simulate nondeterministically M on input x
end

Figure 6.4   Basis for a probabilistic algorithm for a set in NP

---

2 can not accept, because no computation of M accepts x. Hence x is not probabilistically accepted by M''.

If $x \in L$, then some computation of M accepts x; hence, more than half the computations of M'' accept x, namely, all the computations starting with option 1, and some others starting with option 2 and simulating the correct accepting computations of M. Hence x is probabilistically accepted by M''.

Notice that the running time of M'' is just one step more than M.   □

Let us now show some simple properties of the class PP.

Proposition 6.6   (a) *For each set in PP, there is a probabilistic machine that accepts it for which the probability of acceptance is never exactly $1/2$.*
(b) *PP is closed under complementation.*

Proof.

(a) Let M be a probabilistic machine clocked by polynomial p. We can assume that $p(n) > 1$ for all n. Construct a second machine M' which, on inputs of length n, first simulates M for the corresponding $p(n)$ steps, and then tosses $p(n)$ extra coins. With the last move, it accepts if M accepted and at least one of the extra coins came up heads; otherwise it rejects. So in essence each accepting computation of M gets multiplied by $2^{p(n)} - 1$. If M accepts, each of the at least $2^{p(n)-1} + 1$ many accepting computations is so multiplied, and using that $p(n) > 1$ we have more than half (i.e. strictly more than $2^{2p(n)-1}$) accepting computations of M'. On the other hand, if M had at most $2^{p(n)-1}$ many accepting computations, after the multiplication we still have strictly less than $2^{2p(n)-1}$ accepting computations in M'. Therefore M and M' accept the same inputs, and M' never has exactly half the computations accepting.
(b) It is enough to interchange the accepting and rejecting states in the probabilistic Turing machine obtained in Part (a).   □

Corollary 6.7 $NP \cup$ co-$NP \subseteq PP$.

After the first edition of this book, the class PP was shown to be closed under all boolean operations. We only prove here the following closure property:

Proposition 6.8 *PP is closed under symmetric difference.*

Proof. Let A and B be sets in PP. Let M_1 be a probabilistic machine accepting A in polynomial time, and let M_2 be a probabilistic machine accepting B in polynomial time.

Consider the probabilistic machine M of Figure 6.5. This machine works

---

input x
simulate M_1 on x
simulate M_2 on x
if exactly one of M_1 and M_2 accept x
    then accept
    else reject
end

Figure 6.5 Probabilistic algorithm for a symmetric difference

---

in polynomial time. If both M_1 and M_2 produce correct results, then M produces a correct result. Moreover, if both M_1 and M_2 produce incorrect results, then M produces also a *correct* result. Thus, M produces an incorrect result if and only if exactly one of M_1 and M_2 produces an incorrect result.

Hence, if $\frac{1}{2} + \varepsilon$ computations of M_1 are correct, and $\frac{1}{2} + \varepsilon'$ computations of M_2 are correct, then

$$(\frac{1}{2} + \varepsilon) \cdot (\frac{1}{2} + \varepsilon') + (\frac{1}{2} - \varepsilon) \cdot (\frac{1}{2} - \varepsilon') = \frac{1}{2} + 2 \cdot \varepsilon \cdot \varepsilon'$$

computations of M are correct, which is more than $\frac{1}{2}$. Thus M accepts probabilistically the set $A \triangle B$. □

Another closure property of the class PP is the following:

Proposition 6.9 *PP is closed under m-reducibility.*

Proof. Let A be a set in PP, and assume that B is reducible to A via the function f. Let M be a probabilistic machine which accepts probabilistically A in time $p(n)$, where $p(n)$ is a polynomial, and let M' be a machine with output tape which computes f in time $q(n)$, where $q(n)$ is also a polynomial.

Consider a probabilistic machine which works as follows: on input x of length n, it computes $f(x)$, branching into two identical computations at each step, for $q(n)$ steps. Then it simulates M on the result $f(x)$, branching simultaneously with M. This simulation requires $p(q(n))$ steps. The total time taken by this probabilistic machine is $q(n) + p(q(n))$, and it accepts B. Therefore $B \in PP$. □

We finish this section by showing that PP has complete sets. The set MAJ, presented in Example 6.2, is one of them. Another such complete set is defined as follows.

Definition 6.10 *Define the set* #SAT *as follows:*

$$\#\text{SAT} = \{\langle i, F\rangle \mid F \text{ is a boolean formula}$$

$$\text{satisfied by more than } i \text{ assignments }\}$$

The completeness of MAJ and #SAT will follow from the next two lemmas.

Lemma 6.11 MAJ $\in PP$.

Proof. It is enough to check that the probabilistic machine presented in Example 6.2 accepts MAJ in polynomial time, which is quite simple.   □

Lemma 6.12 #SAT *is PP-hard.*

Proof. Let A be a set in PP, and let M be a polynomial time probabilistic machine which accepts A. Let $p(n)$ be the running time of M. Note that x is accepted by M if and only if at least $2^{p(n)-1} + 1$ computations of M accept x, where $n = |x|$.

Now let Accepted(x) be the formula constructed from M as in Theorem 3.23. Observe that the construction of this formula guarantees that each different accepting computation provides a different way of satisfying this formula, and that each satisfying assignment for the formula corresponds to a different accepting computation of M. Therefore, M accepts probabilistically the input x if and only if the formula Accepted(x) has more than $2^{p(n)-1}$ satisfying assignments.

Thus the following function is a polynomial time reduction from A to #SAT:

$$f(x) = \langle 2^{p(n)-1}, \text{Accepted}(x)\rangle$$

Note that $2^{p(n)-1}$ can be written in binary in polynomial time.   □

Now we can easily show the completeness of MAJ and #SAT.

Theorem 6.13 *The sets* MAJ *and* #SAT *are complete for PP under the polynomial time m-reducibility.*

Proof. We must show that both sets are in PP, and that every set in PP is reducible to each of them. We will do this by showing that #SAT is reducible to MAJ.

Let $\langle i, F \rangle$ be an input to #SAT, where F is a formula with m variables. Consider a formula G on the same variables as F, having exactly $2^m - i$ satisfying assignments, as constructed in Example 1.27. Let h be a function defined on pairs $(\langle i, F \rangle)$ as follows: if i is 2^m or more, where m is the number of variables in F, then $h(\langle i, F \rangle)$ is some fixed word not in MAJ. Else, $h(\langle i, F \rangle)$ is the formula $(y \wedge F) \vee (\neg y \wedge G)$ where y is a new boolean variable. Then F has more than i satisfying assignments if and only if $h(\langle i, F \rangle)$ has more than 2^m satisfying assignments, which is half the possible assignments to it (since it has $m + 1$ variables). Therefore $\langle i, F \rangle$ is in #SAT if and only if $h(\langle i, F \rangle)$ is in MAJ. It is easy to compute h in polynomial time.

Thus, #SAT is m-reducible to MAJ. We know also from Lemma 6.12 that #SAT is PP-hard. Hence, by the transitivity of the m-reducibility (see Proposition 3.16), we obtain that MAJ is also PP-hard. But we have shown in Lemma 6.11 that MAJ is in PP. Hence MAJ is PP-complete.

On the other hand, MAJ is in PP and #SAT is reducible to MAJ. By Proposition 6.9, PP is closed under the m-reducibility. Therefore #SAT is also in PP, and Lemma 6.12 indicates that it is PP-hard. Hence #SAT is PP-complete. $\square$

As a closing remark, note that it is easy to find a reduction from MAJ to #SAT: given a formula F with m variables, map it to $g(F) = \langle 2^{m-1}, F \rangle$. Then F is in MAJ if and only if $g(F)$ is in #SAT.

6.4 Bounded Error Probability

Let us turn to the model of a probabilistic machine with bounded error probability. The interest of this model is that it allows us to obtain a result with correct answer, with a probability as high as desired. This is achieved as explained below, by iterating the algorithm as many times as needed. Thus, algorithms of this type are particularly useful, because the probability of error due to the algorithm can be made as small, for example, as the probability of error due to an undetected hardware failure. Moreover, the time cost for reducing the error probability is reasonably low.

Let us start by defining the class BPP (which stands for **B**ounded error **P**robabilistic **P**olynomial time).

Definition 6.14 *BPP is the class of languages recognized by polynomial time probabilistic Turing machines whose error probability is bounded above by some positive constant $\varepsilon < \frac{1}{2}$.*

An equivalent way of expressing this fact is the following: L is in BPP if and only if there is a probabilistic Turing machine M and a positive constant ε such that for every x,

- $x \in L$ if and only if more than $\frac{1}{2} + \varepsilon$ computations of M accept x, and
- $x \notin L$ if and only if more than $\frac{1}{2} + \varepsilon$ computations of M reject x.

From the definition it is clear that $BPP \subseteq PP$. It is not difficult to see that the (non-)primality test of Rabin presented at the beginning of this chapter is of this type: it produces an answer which is erroneous with a probability of at most $(\frac{1}{2})^m$.

Moreover, Rabin's algorithm is guaranteed to obtain a correct answer if the input is a prime. In this case, the probability of error is zero. We present now a class of sets characterized by this property: there is a probabilistic machine accepting it, and the probability of error is bounded for probabilistically accepted inputs, and is zero for probabilistically rejected inputs. It is a kind of "one-sided" or asymmetric version of BPP.

We define it as follows:

Definition 6.15 *R is the class of languages accepted by polynomially clocked probabilistic Turing machines which have zero error probability for inputs not in the language, and error probability bounded by some $\varepsilon < \frac{1}{2}$ for words in the language.*

An equivalent way of expressing this fact is the following: L is in R if and only if there is a probabilistic Turing machine and a positive constant ε such that for every x,

- $x \in L$ if and only if more than $\frac{1}{2} + \varepsilon$ computations of M accept x, and
- $x \notin L$ if and only if every computation of M rejects x.

Any positive constant less than 1 (even greater than $\frac{1}{2}$) can be substituted for the bound on the probability of error in the definition of R (see Exercise 7).

Thus, Rabin's algorithm witnesses the fact that the set of composite numbers is in R. The following theorem locates R.

Theorem 6.16 $P \subseteq R \subseteq NP \cap BPP$.

Proof. P is included in R. To see this, just take a deterministic Turing machine, consider it as a nondeterministic Turing machine, and transform it into a probabilistic machine. On accepted words, all the computation paths accept, while on rejected words all computation paths reject. Therefore $P \subseteq R$.

Of course, $R \subseteq BPP$, since the probability of error is always bounded by ε. To see that $R \subseteq NP$, let L be in R, let M be the probabilistic polynomial time machine witnessing this fact, and look at it simply as a nondeterministic machine. The language it accepts does not change, because for inputs in L the probability of accepting is more than $\frac{1}{2}$ and hence some computation accepts; and for inputs not in L the probability of accepting is zero and hence no computation accepts. Thus M witnesses the fact that L is in NP.

□

Before studying the boolean properties of these classes, we show a construction which allows us to decrease as much as we wish the probability of error of the algorithms "of type" BPP and R.

Theorem 6.17

(a) *A set A is in BPP if and only if for each polynomial p, a polynomial time probabilistic machine can be constructed which accepts A with an error probability of at most $(\frac{1}{2})^{p(|x|)}$.*

(b) *A set A is in R if and only if for each polynomial p, a polynomial time probabilistic machine can be constructed which accepts A with an error probability of at most $(\frac{1}{2})^{p(|x|)}$ on inputs in A, and zero error probability on inputs in $\overline{A}$.*

Proof.

(a) The statement from right to left is immediate. For the converse, let A be a set in BPP, and let M be a probabilistic machine accepting A with error probability $\varepsilon < \frac{1}{2}$. Let δ denote $(1 - \varepsilon)$, the probability of getting a correct answer. Let p be any polynomial. Let $q(n) = c \cdot p(n)$, where c is a constant such that $(4 \cdot \varepsilon \cdot \delta)^c < \frac{1}{2}$. Consider the probabilistic machine M' of Figure 6.6.

```
input x
initialize a counter to 0
for i := 1 to 2 · q(|x|) + 1 do
       simulate M on x
       if M accepts then increment the counter by 1
end for
if the counter is greater than q(|x|)
       then accept
       else reject
end
```

Figure 6.6 Iterating a bounded error probabilistic algorithm

Let $m = 2 \cdot q(|x|) + 1$. The probability that the first j simulations of M give a correct answer is δ^j, and the probability that exactly j fixed simulations of M give a correct answer is $\delta^j \cdot \varepsilon^{m-j}$. More generally, the probability that exactly j (not fixed) simulations of M give a correct answer is $\binom{m}{j} \cdot \delta^j \cdot \varepsilon^{m-j}$.

The probability that M' gives an erroneous answer is the probability that at most $q(|x|)$ simulations give a correct answer, which is:

$$\sum_{j=0}^{q(|x|)} \binom{m}{j} \cdot \delta^j \cdot \varepsilon^{m-j}$$

As $\delta > \varepsilon$, $\delta^j \cdot \varepsilon^{m-j} \leq \delta^{m/2} \cdot \varepsilon^{m/2}$. (Here $m/2$ denotes integer division, so $m/2 = q(|x|)$.) Hence, the error probability of M' is bounded above by

$$\delta^{m/2} \cdot \varepsilon^{m/2} \cdot \sum_{j=0}^{q(|x|)} \binom{m}{j}$$

which is less than $\delta^{m/2} \cdot \varepsilon^{m/2} \cdot 2^{m-1} = (4 \cdot \varepsilon \cdot \delta)^{m/2}$ for $|x|$ large enough. Recalling that m is $2 \cdot q(|x|) + 1$, we have that $m/2$ is $q(|x|) = c \cdot p(|x|)$. Thus the error probability is bounded by $((4 \cdot \varepsilon \cdot \delta)^c)^{p(|x|)}$, which is less than $(\frac{1}{2})^{p(|x|)}$ by the choice of c.

(b) The case of R is similar but easier. The statement from right to left is again immediate. For the converse, let A be a set in R, and let M be a probabilistic machine accepting A with error probability $\varepsilon < \frac{1}{2}$ on accepted inputs and zero error on rejected inputs. Consider the machine M' of Figure 6.7.

```
input x
for i := 1 to p(|x|) do
      simulate M on x
      if M accepts then accept and stop
end for
reject
end
```

Figure 6.7   Iterating a one-sided error probabilistic algorithm

---

If M never accepts then M' never accepts. Assume that x is accepted by M with error probability ε. By construction, M' answers erroneously if and only if all the $q(|x|)$ simulations of M answer erroneously. Thus, the probability of error of M' is $\varepsilon^{p(|x|)}$, which is less than $(\frac{1}{2})^{p(|x|)}$.

Notice that if the error probability ε is greater than $\frac{1}{2}$ but less than 1, a similar argument will work. Just take a constant c such that $\varepsilon^c < \frac{1}{2}$ and loop $c \cdot p(|x|)$ times. See Exercise 7.   □

It follows that the *BPP* and *R* algorithms can be designed so that the error probability can be made as small as desired. We consider this property a decisive argument for considering the sets in *BPP* practically feasible.

Corollary 6.18 *Every language in BPP can be recognized by a polynomial bounded probabilistic Turing machine with error probability smaller than any desired positive constant.*

The boolean properties of *BPP* and *R* are as follows.

Proposition 6.19 *The class BPP is closed under complementation, union and intersection. The class R is closed under union and intersection.*

Proof. That *BPP* is closed under complementation is proved in the same way as in the case of *PP*.

We show the closure under union. Let L_1 and L_2 be languages in *BPP*. Then there exist probabilistic Turing machines M_1 and M_2 such that $L_1 = L(M_1)$ and $L_2 = L(M_2)$ with error probability smaller than any desired positive constant, say $\frac{\varepsilon}{2}$. The probabilistic procedure M_3 described in Figure 6.8 recognizes $L_1 \cup L_2$.

```
input x
test whether x ∈ L(M₁)
test whether x ∈ L(M₂)
accept if at least one of the tests is true
end
```

Figure 6.8   Proving a closure under union

The error probability of M_3 is at worst $\frac{\varepsilon}{2} + \frac{\varepsilon}{2} = \varepsilon$. The closure of *BPP* under intersection follows from the closure under union and complementation.

The same algorithm will work for the case of *R*, and the analogous algorithm which checks that both tests are true shows the closure of *R* under intersection.   □

It is not known whether *R* is closed under complementation. Following our notation, co-*R* denotes the class of languages *L* such that $\overline{L} \in R$. From the closure of *BPP* under complementation, together with Theorem 6.16, we can state the following:

Proposition 6.20   co-$R \subseteq BPP$.

Other closure properties of *R* and *BPP* are given in Exercise 9.

6.5 Nonuniform Properties of *BPP*

In this section we present a very interesting property that characterizes the sets in *BPP*. It can be stated in terms of the advice classes used in Chapter 5 to define nonuniform complexity classes. This property is a consequence of Theorem 6.17, and allows us to prove that every set in the class *BPP* has polynomial size boolean circuits. Moreover, the exact relationship of the class *BPP* with the polynomial time hierarchy, to be defined in Chapter 8, relies heavily on this characterization. Still another very interesting application will be presented in Volume II.

In order to simplify the statement of the characterization, we define the "correct advices" for a set at a given length, as follows.

Definition 6.21   *Fix a set $B \in P$. Given a set A and an integer n, the word y is a correct advice for A at length n if and only if*

$$\forall x, |x| = n, (x \in A \text{ if and only if } \langle x, y \rangle \in B)$$

The characterization may be interpreted as: "the sets in *BPP* are those sets that have many short correct advices for each length". Formally, this is stated as follows.

Theorem 6.22   *The following are equivalent:*

(a)   *$A \in BPP$.*
(b)   *Let q be any polynomial. Then there are a set $B \in P$ and a polynomial p such that for each length n the following holds: among all the words of length $p(n)$ there are at least*

$$2^{p(n)} \cdot \left(1 - \left(\frac{1}{2} \right)^{q(n)} \right)$$

   correct advices for A at length n.

(Note that statement (b) can be interpreted as follows: the ratio of *incorrect* advices to the total number of words of length $p(n)$ is at most $(\frac{1}{2})^{q(n)}$.)
Proof.   Let us prove that (b) implies (a). Take any polynomial q such that $q(n)$ is always greater than 2, and let B and p be as indicated. Consider the nondeterministic machine M of Figure 6.9.

Convert it into a probabilistic machine. If the guessed word is a correct advice for A at length $|x|$, then it decides correctly whether x is in A. Each computation of M corresponds to a word y of length $p(|x|)$. Thus the error probability is the ratio of incorrect advices to the total number of words of length $p(|x|)$, which is less than $(\frac{1}{2})^{q(|x|)}$. As q is always greater that 2, the error probability is bounded by $\frac{1}{4}$. Thus M witnesses the fact that A is in *BPP*.

```
input x
guess y of length p(|x|)
if ⟨x, y⟩ ∈ B then accept
    else reject
end
```

Figure 6.9 A nondeterministic algorithm

For the converse, let A be a set in *BPP*, and let q be any polynomial. Using Theorem 6.17, obtain a probabilistic machine M which accepts A with an error probability bounded above by $(\frac{1}{2})^{q(n)+n}$. Denote by p the polynomial running time of M. Each word of length $p(n)$ can be associated to a computation of M on a word of length n, as in the proof of Theorem 6.3.

Let us count the number of words of length $p(n)$ corresponding to erroneous computations. Fix any x of length n. Then the bound on the error probability of M implies that at most $2^{p(n)-q(n)-n}$ words of length $p(n)$ correspond to computations which are incorrect on x. Therefore, at most $2^n \cdot 2^{p(n)-q(n)-n}$ words of length $p(n)$ correspond to computations which are incorrect on some x of length n. All the remaining words of length $p(n)$ must correspond to computations that are correct on *every* input word x of length n.

Now define the following set:

$$B = \{\langle x, y\rangle \mid |y| = p(|x|), \text{ and the computation of } M$$

$$\text{corresponding to } y \text{ on input } x \text{ accepts } \}$$

It is clear that B can be decided in polynomial time. For this B, and the polynomial p indicated above, all the words of length $p(n)$ corresponding to computations of M which answer correctly on every input of length n are, by definition, correct advices for A at this length, and there are

$$2^{p(n)} - 2^n \cdot 2^{p(n)-q(n)-n} = 2^{p(n)} \cdot \left(1 - \left(\frac{1}{2}\right)^{q(n)}\right)$$

of them, as was to be shown. This completes the proof.   □

As a consequence, we have shown that sets in *BPP* can be decided in polynomial time with the help of a polynomially long advice; moreover, "many" words of length $p(n)$ are admissible as values of the advice. We can state this fact as follows:

Corollary 6.23   *If A is a set in BPP then A has polynomial size circuits.*

Proof. Let A be a set in *BPP*. Consider any polynomial q and construct the set B and the polynomial p as in Theorem 6.22. We know that for each length n there is at least one word of length $p(n)$ which is a correct advice for A at length n. Define the function h mapping each word of the form 0^n to the first word of length $p(n)$ which is such a correct advice. Then, if $|x| = n$,

$$x \in A \text{ if and only if } \langle x, h(0^n) \rangle \in B$$

and hence A is in *P/poly*. Therefore, by Theorem 5.25, A can be decided by polynomial size circuits.   □

From the inclusion relationships among the classes R, co-R and *BPP* proven in the previous section, we conclude that:

Corollary 6.24   *Sets in R and sets in co-R can be decided by polynomial size circuits.*

6.6 Zero Error Probability

Algorithms in which random or pseudo-random decisions are made are called "Monte-Carlo" algorithms. In principle, these algorithms may give incorrect answers with a certain probability. The uses of Monte-Carlo algorithms include numerical analysis applications, such as approximate resolution of integration problems. From the complexity-theoretic point of view, these algorithms can be identified by the following characteristics: they are efficient, or at least feasible, and may give incorrect answers. Hence, the class *PP* is usually admitted as a reasonable formalization of the concept of Monte-Carlo algorithm. The classes *BPP* and R may be thought of as more restrictive formalizations of this notion of Monte-Carlo algorithm.

We present in this section a subclass of the probabilistic algorithms characterized by: efficiency—or at least feasibility—, absolute reliability of the correctness of the answer, and a probability that the algorithm halts without answer. Again this probability should be bounded, so that iterating the algorithm allows one to increase as much as desired the probability of getting an answer. Probabilistic algorithms of this kind are known as "Las Vegas" algorithms.

We continue to identify feasibility with polynomial running time. To give the algorithm the chance of halting without answer, we assume during this section that probabilistic machines have three final states, interpreted as ACCEPT, REJECT, and "?" ("I don't know"). We will call them 3-output probabilistic machines. For 3-output probabilistic machines, acceptance is defined as for our usual probabilistic machines: more than half the computations halt in the ACCEPT final state. The error probability for 3-output machines is defined as follows.

Definition 6.25 *The error probability of a 3-output machine is the probability of halting in the REJECT final state on an accepted input, and the probability of halting in the ACCEPT final state on a rejected input.*

Now, the Las Vegas algorithms can be identified with the probabilistic class *ZPP* defined as follows.

Definition 6.26 *ZPP is the class of languages recognized by polynomial time 3-output probabilistic Turing machines with zero error probability.*

This means that on accepted inputs, more than half the computations accept and no computation rejects (the others answer "?"), while on rejected inputs more than half the computations reject and no computation accepts.

In the same way that we proved Proposition 6.19, we can prove the following (Exercise 11).

Proposition 6.27 *The class ZPP is closed under the operations of complementation, union, and intersection.*

The class *ZPP* is related to the other probabilistic complexity classes, and in particular to R, as follows:

Theorem 6.28 $ZPP = R \cap \text{co-}R$.

Proof. *ZPP* is included in R, because the *ZPP* machine can be transformed into an R machine just by identifying the REJECT and the "?" final states. As *ZPP* is closed under complementation, *ZPP* is also included in co-R.

To prove $R \cap \text{co-}R \subseteq ZPP$, let L and $\overline{L}$ both be in R. Then there are "R" machines M_1 and M_2 such that $L = L(M_1)$ and $\overline{L} = L(M_2)$.

Let us consider an algorithm to decide whether $x \in L$; see Figure 6.10.

```
input x
simulate M₁ on x
simulate M₂ on x
if M₁ accepts x then accept
    else if M₂ accepts x then reject
    else halt in the "?" final state
end
```

Figure 6.10 A zero error probabilistic algorithm

The error probability of this algorithm is 0, because by definition of R, if x is accepted by M_1 then it is certain that $x \in L$, and if x is accepted by M_2 then it is certain that $x \in \overline{L}$. The probability of getting a "?" answer is the probability that both machines reject; since every word is either in L or in $\overline{L}$, if both machines reject then one of them is making an error. By the definition of R, the probability that this happens is less than $\frac{1}{2}$. So $L \in ZPP$.   □

This characterization allows us to compare ZPP with the deterministic and nondeterministic classes.

Corollary 6.29 $P \subseteq ZPP \subseteq NP \cap co\text{-}NP$.

Proof. As $P \subseteq R \subseteq NP$, $P \subseteq R \cap co\text{-}R \subseteq NP \cap co\text{-}NP$. The statement follows from Theorem 6.28.   □

From this characterization, together with Corollary 6.24, we can state the following:

Corollary 6.30   *Sets in ZPP can be decided by polynomial size circuits.*

The diagram of Figure 6.11 summarizes the known relationships between the polynomial time deterministic, nondeterministic, and probabilistic classes. Each arrow stands for an inclusion relationship.

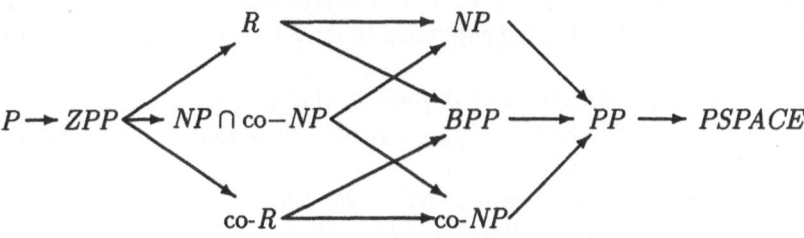

Figure 6.11   Relationships between probabilistic classes

None of the above inclusions is known to be proper. Usually, BPP is the broader of the classes in the diagram which is admitted to consist of computationally feasible problems. This is because the iteration of the algorithm allows one to obtain answers within a reasonably small amount of time and with a reasonably high probability that the answer is correct.

6.7 Exercises

1.   •Find an alternative proof of Proposition 6.6(a). Show that any other constant fraction greater than 0 can be substituted for the $\frac{1}{2}$ in the definition of PP. Show also that the same class can be defined requiring at least some fixed (integer) number of accepting computations (instead

of a fraction of the total) to consider an input accepted. *Hint:* For the first part, first ensure to get an even number of accepting computations by an extra coin toss, then start over with a coin toss to go into the simulation or to accept with a fixed odd probability, so that the total probability is always odd.

2. The class D^P is formed by all sets which can be expressed as the intersection of a set in NP and another set in co-NP. Show that D^P is included in PP, using the closure of PP under symmetric difference.

3. Use the proof of the last exercise to show that the boolean closure of NP is included in PP.

4. Prove that a language L is in PP if and only if it can be defined as follows: $x \in L$ if and only if $\langle x, y \rangle \in Q$ for more than half of the words y with $|y| \leq p(|x|)$, where p is a polynomial and Q is in P.

5. Prove that a language L is in BPP if and only if it can be defined as follows: $x \in L$ if and only if $\langle x, y \rangle \in Q$ for more than $\frac{3}{4}$ of the words y with $|y| \leq p(|x|)$, and $x \notin L$ if and only if $\langle x, y \rangle \notin Q$ for more than $\frac{3}{4}$ of the words y with $|y| \leq p(|x|)$, where p is a polynomial and Q is in P.

6. Show that #SAT is self-reducible. *Hint:* On input $\langle i, F \rangle$, use binary search and the oracle #SAT to find in polynomial time a lower bound on the number of satisfying assignments for $F|_{x:=0}$ and $F|_{x:=1}$, where x is a variable of F.

7. Show that in the definition of R given in the text, the condition $\varepsilon < \frac{1}{2}$ is unnecessary and can be dropped. *Hint:* Check that $\varepsilon < 1$ is enough to prove Theorem 6.17, part (b).

8. •Show that $NP \subseteq BPP$ implies $NP = R$. *Hint:* Construct a machine "of type R" for SAT as follows: use the BPP machine for SAT to look for a satisfying assignment, as in the proof of Theorem 5.28. Accept only if the assignment found satisfies the input.

9. Show that the classes ZPP, R, and BPP are closed under polynomial time m-reducibility. Which of these classes are closed under polynomial time T-reducibility? Why? *Hint:* Take a look at Exercise 4 of Chapter 4.

10. •Recall the definition of strong nondeterministic Turing machines from Exercise 24 of Chapter 4. Which conditions are to be imposed in order to define ZPP using these machines? Express these conditions in such a way that one is obtained from the other by a swap of quantifiers. Use the same swap of quantifiers to obtain a description of R from a description of NP, and a description of BPP from a description of PP. Discuss the intuitive meaning of these descriptions.

11. Prove Proposition 6.27.

12. Define relativizations of the probabilistic polynomial time complexity classes. For each of the results in this chapter, argue whether it holds in the relativized case.

13. Prove that if $P \neq R$ then $DEXT \neq EXPSPACE$. *Hint:* Use Exercise 17 of Chapter 5.

14. *Prove that $P_{help}(BPP) \subseteq ZPP$. (See Exercise 33 of Chapter 4 for the definition of P_{help}.)

6.8 Bibliographical Remarks

The complexity-theoretic study of probabilistic algorithms which increase the speed of resolution of some problems dates back to 1976, with the algorithm to test primality in polynomial time described in the text. This algorithm is due to Rabin (1976), and is related to the work of Miller (1976). At the same time and independently, Solovay and Strassen (1977) found another similar probabilistic algorithm to test primality—see also Solovay and Strassen (1978). Since then, several probabilistic algorithms have been proposed for solving a variety of problems in Bach, Miller, and Shallit (1984), Reif (1985), and others. A very good survey is Welsh (1983), where more motivation for the consideration of these algorithms is also given. A very recent result in the area, announced by Adleman and Huang (1987), is a polynomial time Las Vegas algorithm for primality. Thus, PRIMES $\in ZPP$.

The first model of probabilistic Turing machine goes back to de Leeuw, Moore, Shannon, and Shapiro (1956). An appropriate model for the probabilistic algorithms, the probabilistic Turing machine, was fully developed by J. Gill in his Ph. D. dissertation. He defined our polynomial time probabilistic classes (see Gill (1972 and 1977)).

According to Gill, a probabilistic Turing machine is a Turing machine with distinguished states called *coin-tossing* states. For each coin-tossing state, the finite control unit specifies two possible next states. The computation of a probabilistic Turing machine is deterministic except that in coin-tossing states the machine tosses an unbiased coin to decide between the next two possible states. In Gill's model of probabilistic Turing machine, acceptance is defined as follows: we say that machine M *accepts* the input x if the probability of M accepting x is strictly greater than $\frac{1}{2}$. He also considers probabilistic Turing machines with output, which are able to compute functions.

This model can be extended by relaxing the requirement that unbiased random decisions be made. This approach was studied in Santos (1969 and 1971). Another closely related concept that we have not developed here is the notion of "counting function" associated to a nondeterministic machine, and the corresponding important complexity class $\#P$, defined in Valiant (1979).

Using his model, Gill develops most of the standard recursion theory, proving that the functions computed by probabilistic machines are exactly the partial recursive functions. He also defined PP, BPP, R (called VPP in Gill (1977)) and ZPP, and demonstrated many properties of these classes. The completeness of MAJ and of #SAT are shown in the text by combining arguments of Gill (1977) and of Simon (1975). Exercise 1 is also from this last reference, where it is presented in the framework of the "threshold machines".

The model presented in the text is a slight modification of the model proposed by Gill, although it is easily seen that they give equivalent definitions of the polynomial time classes. Our model is from Russo (1985), who proves also the closure of PP under symmetric difference. Our presentation of Theorem 6.17—from Zachos (1982)—also follows this reference. We have obtained from Russo (1985) Exercise 2—this fact was shown in Papadimitriou and Yannakakis (1982) and Papadimitriou and Zachos (1985)—and Exercise 3. After the first edition of this volume, Beigel, Reingold, and Spielman (1994) have shown that PP is closed under all boolean operations, and Fortnow and Reingold (1991) proved that it is closed under polynomial time truth-table reducibility.

The fact that R has polynomial size circuits was shown by Adleman (1978). He also coined the name R (other authors use RP). Our presentation, via the characterization of BPP, follows Schöning (1985a). Basic ideas of this construction and of some of the exercises appear also in Bennett and Gill (1981), Hinman and Zachos (1984), and Zachos and Heller (1986).

Exercise 6 is from Balcázar, Book, and Schöning (1986), and Exercise 8 is from Ko (1982). The relativizations of probabilistic classes, and some interesting properties that we will present in Volume II, appear in Rackoff (1982). Exercise 14 is from Schöning (1985a), and the converse inclusion is still an open problem at the time of writing.

Finally, it should be indicated that in Lautemann (1982) other models of probabilistic computation are proposed and compared with the probabilistic Turing machine.

7   Uniform Diagonalization

7.1 Introduction

This chapter presents a powerful technique for proving the existence of certain types of "diagonal" recursive sets: the Uniform Diagonalization Theorem. It allows one to prove the existence of non-complete sets in $NP - P$, provided that $P \neq NP$. We will show also, using this theorem, that under the same hypothesis infinite hierarchies of incomparable sets (with respect to polynomial time reducibilities) exist in $NP - P$. This theorem allows the original proofs of these results to be considerably simplified, and we will use it later to translate the results to other complexity classes.

7.2 Presentability and Other Properties

We start with some basic definitions and lemmas.

Definition 7.1   *A class $\mathcal{C}$ of sets is recursively presentable if and only if there is an effective enumeration $M_1, M_2, \ldots$ of deterministic Turing machines which halt on all their inputs, and such that $\mathcal{C} = \{L(M_i) \mid i = 1, 2, \ldots\}$.*

By convention, we consider the empty class to be recursively presentable. Notice that some of the sets of a recursively presentable class $\mathcal{C}$ (in most cases, all the sets) may appear several times (in most cases, infinitely many times) in the enumeration.

Observe that the definition only makes sense for classes of sets which are strictly included in the class of recursive sets. It is left to the reader (Exercise 3) to prove that the class of recursive sets is not recursively presentable.

Definition 7.2   *Let $r : \mathbb{N} \to \mathbb{N}$ be a recursive function such that $r(m) > m$ for all m. Fix an alphabet Γ. Define the set $G[r]$ as:*

$$G[r] = \{x \in \Gamma^* \mid r^n(0) \leq |x| < r^{n+1}(0) \text{ for } n \text{ even}\}$$

where $r^n(m)$ denotes the n-fold application of r to m:

$$\overbrace{r \cdot r \cdot r \cdots r}^{n}(m)$$

Thus, $G[r]$ is formed by the words in the shadowed pieces of Figure 7.1. Sometimes, $G[r]$ has been called the *gap language* generated by r.

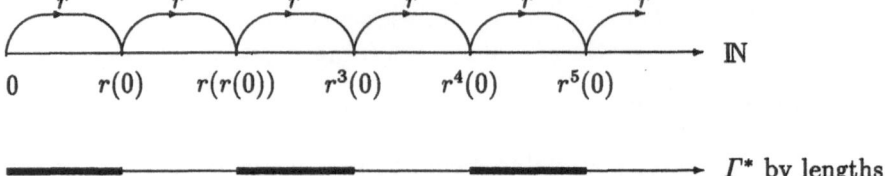

Figure 7.1   The *gap language* generated by r

Under the hypothesis that r is time constructible, we can show that it is not difficult to decide membership of $G[r]$.

Proposition 7.3   *If r is time constructible, then $G[r] \in P$.*

Proof.   On input x, we compute (in unary) $0, r(0), r(r(0)), \ldots$ until a word of length $|x|$ is reached; let n be such that $r^n(0) \leq |x| < r^{n+1}(0)$, and accept if and only if n is even. Notice that it is not necessary to compute $r^{n+1}(0)$; it suffices to spend $|x| + 1$ steps in computing it. By the time constructibility of r, this ensures that it is greater than $|x|$. By the same reason, each computation of $r^i(0)$ only needs about $r^i(0) < |x|$ steps. On the other hand, we will only perform a maximum of $|x|$ applications of r. Therefore, the running time of this procedure is bounded by $|x|^2$.   □

The fundamental property of the recursively presentable classes is that we can use their presentation to diagonalize over them, constructing sets that lie outside the class. We present in the next section a theorem which constructs diagonal sets in an "automatic" way. Other diagonalizations to be presented in the remaining part of this book will also use recursive presentations of the classes over which the diagonalization is performed.

7.3 The Main Theorem

The following result is known as the *Uniform Diagonalization Theorem*.

Theorem 7.4   *Let A_1, A_2 be recursive sets and $\mathcal{C}_1$, $\mathcal{C}_2$ be classes of recursive sets with the following properties:*

1. $A_1 \notin \mathcal{C}_1$
2. $A_2 \notin \mathcal{C}_2$
3. $\mathcal{C}_1$ and $\mathcal{C}_2$ are recursively presentable.
4. $\mathcal{C}_1$ and $\mathcal{C}_2$ are closed under finite variations.

Then there exists a recursive set A such that

(a) $A \notin \mathcal{C}_1$
(b) $A \notin \mathcal{C}_2$
(c) $A \leq_m A_1 \oplus A_2$

Proof. Let $P_1, P_2, \ldots$ and $Q_1, Q_2, \ldots$ be effective enumerations of Turing machines that present $\mathcal{C}_1$ and $\mathcal{C}_2$ respectively. Define the following functions:

$$r_1(n) = \max_{i \leq n}\{|z_{i,n}|\} + 1$$

$$r_2(n) = \max_{i \leq n}\{|z'_{i,n}|\} + 1$$

where $z_{i,n}$ is the smallest word in Γ^* such that $|z_{i,n}| > n$ and $z_{i,n} \in (L(P_i) \triangle A_1)$, while $z'_{i,n}$ is the smallest word in Γ^* such that $|z'_{i,n}| > n$ and $z'_{i,n} \in (L(Q_i) \triangle A_2)$.

Notice that $z_{i,n}$ and $z'_{i,n}$ always exist. Otherwise, if there were no words of length greater than n in $L(P_i) \triangle A_1$ (respectively, in $L(Q_i) \triangle A_2$), then $L(P_i)$ would be a finite variation of A_1 (respectively, $L(Q_i)$ of A_2), and since $\mathcal{C}_1$ and $\mathcal{C}_2$ are closed under finite variations, we get $A_1 \in \mathcal{C}_1$ (respectively $A_2 \in \mathcal{C}_2$), which is a contradiction to the hypothesis of the theorem.

For any n and every $i \leq n$, there is a recursive procedure for finding $z_{i,n}$ in $(L(P_i) \triangle A_1)$ and $z'_{i,n}$ in $(L(Q_i) \triangle A_2)$; this is because, as P_i and Q_i are Turing machines which always halt, $L(P_i)$ and $L(Q_i)$ are recursive sets, and as A_1 and A_2 are also recursive sets, $(L(P_i) \triangle A_2)$ and $(L(Q_i) \triangle A_2)$ are recursive sets. To find the maximum and to add 1 is also a recursive procedure, so we have proved that r_1 and r_2 are total recursive functions.

Now let $r \geq \max(r_1, r_2)$ be a nondecreasing time constructible function, which exists by Lemma 2.23. For every n, by the construction of r_1, the words $z_{i,n}$, which are the counterexamples showing that $A_1 \neq L(P_i)$, are located between n and $r_1(n)$; and by construction of r_2, the counterexamples $z'_{i,n}$ showing $A_2 \neq L(Q_i)$ are between n and $r_2(n)$. Therefore, for every n, there are between n and $r(n)$ some counterexamples witnessing the fact that $\forall i \leq n, A_1 \neq L(P_i)$ and $A_2 \neq L(Q_i)$. See Figure 7.2. Define the set $A = (G[r] \cap A_1) \cup (\overline{G[r]} \cap A_2)$. Notice that the definition forces the coincidence of A with A_1 in the "even jumps" of r, and with A_2 in the "odd jumps" of r.

We shall prove that A has the desired properties.

(a) $A \notin \mathcal{C}_1$. For suppose that $A \in \mathcal{C}_1$, then there exists i such that $A = L(P_i)$. Let m be an even integer such that $r^m(0) \geq i$. From the

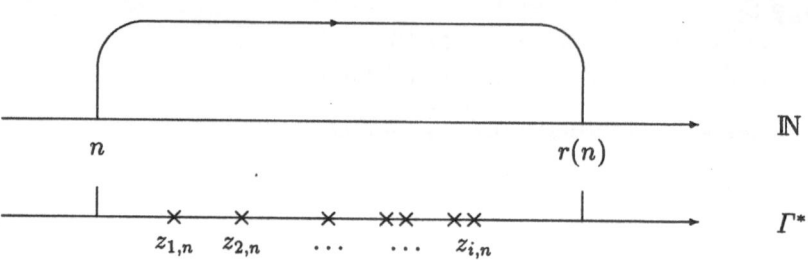

Figure 7.2 "Counterexamples" between n and $r(n)$

construction of r it follows that there exist words $z \in \Gamma^*$ with $r^m(0) \leq |z| < r^{m+1}(0)$ and such that $z \in (L(P_i) \triangle A_1)$. Notice that as m is even, z will be also in $G[r]$. But as A coincides with A_1 in $G[r]$, it follows that $z \in (L(P_i) \triangle A)$, which is a contradiction to the hypothesis that $A = L(P_i)$. Therefore $A \notin C_1$.

(b) $A \notin C_2$. This is proved by an argument completely symmetric to part (a).

(c) $A \leq_m A_1 \oplus A_2$. As $G[r] \in P$, the function defined by:

$$f(x) = \begin{cases} x0 & \text{if } x \in G[r] \\ x1 & \text{if } x \notin G[r] \end{cases}$$

is a polynomial time reduction from A to $A_1 \oplus A_2$.   □

7.4 Applications

This theorem can be applied in a broad variety of forms. We present some of the main applications in this section, and we leave others as part of the exercises at the end of this chapter. We will present other applications later.

In order to apply this theorem we must show that the classes we use are recursively presentable. Thus, we begin by proving some lemmas that allow us to show the presentability of most complexity classes.

Lemma 7.5 *For every recursive set A, the class $P(A)$ is recursively presentable.*

Proof. Let $M_1, M_2, \ldots$ be an enumeration of all the deterministic oracle Turing machines. We construct a new enumeration by attaching "clocks" to the machines that stop them after a polynomial number of steps. This is done as follows: for each pair of natural numbers i, j, consider the machine

$P_{\langle i,j \rangle}$ which on input x simulates $|x|^j$ steps of the computation of M_i on x, using A as oracle. As $P_{\langle i,j \rangle}$ works in polynomial time, the set that it accepts is in $P(A)$. Furthermore, every set in $P(A)$ is accepted by some M_i within some time bound n^j, and hence it is accepted by some machine $P_{\langle i,j \rangle}$. The enumeration of all $P_{\langle i,j \rangle}$ is effective, and all the machines halt for every input. The machines in this enumeration are called "polynomially clocked" oracle Turing machines. Observe that the property that every polynomial is time constructible is used here (see Example 2.22).

As A is recursive, there is a machine M which accepts A and halts for every input. Now substitute calls to M for the oracle queries of the machines $P_{\langle i,j \rangle}$, obtaining a new effective enumeration $Q_{\langle i,j \rangle}$. By definition, this is a presentation of $P(A)$. Therefore $P(A)$ is recursively presentable.   □

Notice that clocking Turing machines with output tape in the same manner produces a similar enumeration of machines presenting all the partial functions computable in polynomial time. This fact will be used later.

Lemma 7.6 *If C is recursively presentable, then co-C is also recursively presentable.*

Proof. Just exchange the accepting and rejecting states of the machines in a presentation of C to obtain a presentation of co-C.   □

From now on, denote by $\leq_r$ any of the two polynomial time reducibilities $\leq_m$ or $\leq_T$. The next results show a way to obtain recursively presentable subclasses of recursively presentable classes.

Lemma 7.7 *Let C_1 and C_2 be two recursively presentable classes, and let A be any recursive set. Assume that C_1 is closed under finite variations. Then the following classes are recursively presentable:*

(a) $C_3 = \{B \in C_1 \mid \exists D \in C_2 \text{ such that } B \oplus A \leq_r D\}$

(b) $C_4 = \{B \in C_1 \mid \exists D \in C_2 \text{ such that } D \leq_r B \oplus A\}$

Proof. We prove the presentability of C_3 for the case of the polynomial time Turing reducibility.

If C_3 is empty then it is recursively presentable by convention. Assume that C_3 is non-empty, and let E be a set in C_3. From the definition of C_3 and the closure of C_1 under finite variations, it follows that all the finite variations of E are also in C_3.

Let $P_1, P_2, \ldots$ be a presentation of C_1, and let $Q_1, Q_2, \ldots$ be a presentation of C_2. Let $T_1, T_2, \ldots$ be an enumeration of deterministic oracle Turing machines with polynomial clocks as in Lemma 7.5.

We construct an enumeration of machines for C_3. For $n = \langle i, j, k \rangle$, let M_n be the machine that operates as indicated in Figure 7.3.

Machine M_n:

input x

for each y with $|y| < |x|$ do

    test that $y \in (L(P_j) \oplus A)$ if and only if $y \in L(T_k, L(Q_i))$

if all tests are true

    then accept x if and only if $x \in L(P_j)$

    else accept x if and only if $x \in E$

Figure 7.3 Showing recursive presentability

If T_k computes a reduction from $L(P_j) \oplus A$ to $L(Q_i)$, then $L(P_j) = L(M_n)$ for $n = \langle i, j, k \rangle$, which implies that every set in the class C_3 is accepted by some M_n. On the other hand, if T_k does not compute such a reduction, then $L(M_n)$ is a finite variation of E and therefore belongs to C_3. Therefore the machines M_n form a presentation of C_3.

The case of the m-reducibility and the presentability of C_4 for both reducibilities are proven in an analogous manner (see Exercise 2). □

These three lemmas allow us to prove the presentability of a number of classes. Lemma 7.7 will be used in its full generality only in one place, in the proof of Theorem 7.10 below. Usual applications of Lemma 7.7 will employ a particular case, with $A = \emptyset$ and with C_2 consisting of only one set.

Proposition 7.8 *The following classes are recursively presentable:*

(a) *Any finite class of recursive sets.*

(b) *PSPACE.*

(c) *NP.*

(d) *co-NP.*

(e) *The class of the NP-complete sets with respect to polynomial time m- and T- reducibilities.*

(f) *The class of the co-NP-complete sets with respect to polynomial time m- and T- reducibilities.*

(g) *The class of the PSPACE-complete sets with respect to polynomial time m- and T- reducibilities.*

(h) *The relativizations of any of the above classes to any fixed recursive set.*

Proof.

(a) Let $M_1, M_2, \ldots M_n$ be machines accepting the finitely many sets in the class. Define $M_k = M_n$ for $k \geq n$. It is an effective enumeration because it is ultimately constant, so by definition the class is recursively presentable.

(b)   Recall from Theorem 3.29 that QBF is *PSPACE*-complete. By the closure of *PSPACE* under T-reducibility, we have that $PSPACE = P(QBF)$. Thus, by Lemma 7.5, *PSPACE* is recursively presentable.

(c)   *NP* can be defined as $\{B \in PSPACE \mid B \leq_m SAT\}$. Therefore, taking $C_1 = PSPACE$, $C_2 = \{SAT\}$, and $A = \emptyset$, *NP* is recursively presentable by part (a) of Lemma 7.7. Notice that this part cannot be shown directly by clocking nondeterministic machines as in Lemma 7.5, since the definition of recursive presentation requires deterministic machines.

(d)   Follows from the presentability of *NP* and Lemma 7.2.

(e)   The *NP*-complete sets are defined as $\{B \in NP \mid SAT \leq_m B\}$. Thus it is recursively presentable by part (b) of Lemma 7.7, taking $C_1 = NP$, $C_2 = \{SAT\}$, and $A = \emptyset$.

(f)   Follows from the presentability of the *NP*-complete sets and Lemma 7.6.

(g)   It is proved in a manner analogous to the case of the *NP*-complete sets.

(h)   The statement follows from the fact that all the proofs presented up to now relativize easily.   □

Now we can prove the existence of non-complete sets in $NP - P$, under the hypothesis that $P \neq NP$.

Theorem 7.9   *If $P \neq NP$, then there are sets in NP which are neither in P nor NP-complete.*

Proof.   Let $C_1 = P$, C_2 the *NP*-complete sets, $A_1 = SAT$ and $A_2 = \emptyset$. Under the hypothesis $P \neq NP$, $A_1 \notin C_1$ and $A_2 \notin C_2$. Both classes are recursively presentable and closed under finite variations. By the Uniform Diagonalization Theorem, there exists an A such that $A \notin C_1$ and $A \notin C_2$, but $A \leq_m SAT \oplus \emptyset$. By the closure of *NP* under the m-reducibility, A is in *NP*. Thus A is the desired set.   □

We will extend this theorem to the construction of infinite families of incomparable sets in $NP - P$. Given a reducibility, we say that sets A and B are incomparable if neither one is reducible to the other.

Theorem 7.10   *Let A and B be recursive sets such that $A \leq_T B$ but $B \not\leq_T A$. Then there exists an infinite family of sets D_i, $i \in \mathbb{N}$, such that:*

(a)   *For all i, $A \leq_T D_i \leq_T B$ but $B \not\leq_T D_i \not\leq_T A$.*

(b)   *For every i and j, if $i \neq j$ then D_i and D_j are incomparable.*

(c)   *Moreover, if $A \leq_m B$ then, for all i, $A \leq_m D_i \leq_m B$ but $B \not\leq_T D_i \not\leq_T A$.*

Proof.   Let $C = P(B)$. By Lemma 7.5, C is recursively presentable. We construct the sets $\{D_i\}$ inductively.

Case $n = 0$. Consider the following two classes: $\mathcal{C}_1 = \{X \in P(B) \mid X \leq_T A\}$ and $\mathcal{C}_2 = \{X \in P(B) \mid B \leq_T X \oplus A\}$. As $\emptyset \in \mathcal{C}_1$ and $B \in \mathcal{C}_2$, both classes are nonempty, and by Lemma 7.7 both are recursively presentable classes. Moreover, as $A \leq_T B$, then $B \notin \mathcal{C}_1$ and $\emptyset \notin \mathcal{C}_2$. Therefore, from the Uniform Diagonalization Theorem we obtain a set E such that $E \leq_m B \oplus \emptyset$ (which implies $E \leq_T B \oplus \emptyset$), and hence E is in $P(B)$, but $E \notin \mathcal{C}_1$ and $E \notin \mathcal{C}_2$.

From the definition of $\mathcal{C}_1$ and $\mathcal{C}_2$, it follows now that $E \not\leq_T A$, which implies that $E \oplus A \not\leq_T A$, and that $B \not\leq_T E \oplus A$. Define $D_0 = E \oplus A$: then (a) and (c) hold for D_0. Part (b) will be ensured by the induction step.

From n to $n+1$. Assume that we have already defined the sets $D_0, D_1, \ldots, D_n$ satisfying the hypotheses (a), (b), and (c). Let

$$\mathcal{C}_1 = \{X \in P(B) \mid \exists i \, X \leq_T D_i\} \text{ and } \mathcal{C}_2 = \{X \in P(B) \mid \exists i \, D_i \leq_T X \oplus A\}$$

Then $\emptyset \in \mathcal{C}_1$ and $B \in \mathcal{C}_2$, and by Lemma 7.7 both are recursively presentable. Moreover, $B \notin \mathcal{C}_1$ and $\emptyset \notin \mathcal{C}_2$ by part (a) of the induction hypothesis. Therefore, using the Uniform Diagonalization Theorem, there exists a set E such that $E \leq_m B \oplus \emptyset$, and hence $E \leq_T B \oplus \emptyset$. Now, $B \oplus \emptyset$ is m-equivalent to B, and hence E is in $P(B)$, but $E \notin \mathcal{C}_1$ and $E \notin \mathcal{C}_2$. From the definition of $\mathcal{C}_1$, it follows that for every D_i, $E \not\leq_T D_i$ which in turn implies that for every D_i, $E \oplus A \not\leq_T D_i$. From the definition of $\mathcal{C}_2$, it follows that for every $i \leq n$, $D_i \not\leq_T E \oplus A$.

Defining $D_{n+1} = E \oplus A$, we have that D_{n+1} is not comparable with any other D_i with $i \leq n$. Incomparability with the remaining sets D_i will be ensured by later inductive steps. This shows (b). Property (c) is immediate. It remains to show (a). We know that A is T-reducible to D_{n+1} and that D_{n+1} is T-reducible to B. Fix any set D_i with $i \leq n$. If D_{n+1} is T-reducible to A, then by transitivity it is T-reducible to D_i, contradicting the incomparability; and if B is T-reducible to D_{n+1}, then by transitivity D_i is T-reducible to D_{n+1}, contradicting again the incomparability. Thus (a) holds for D_{n+1}. □

Corollary 7.11 *If $P \neq NP$, then there exist infinitely many incomparable T-degrees in NP.*

Proof. Use the last theorem with $A = \emptyset$ and $B = \text{SAT}$. Notice that $A \leq_m B$, and therefore part (c) of Theorem 7.10 guarantees that all the constructed sets remain in NP. □

The last two results have been stated for the T-reducibility. Analogous results can be established for the m-degrees. See Exercise 14.

Some more consequences of the Uniform Diagonalization Theorem are presented in the exercises, and some others will be found in the next chapter and later in Volume II.

7.5 Exercises

1. Show that the class of all finite sets is recursively presentable.
2. Complete the proof of Lemma 7.7.
3. Prove that the following classes are *not* recursively presentable:

    (a) The class of all the recursive sets.
    (b) The class of the sparse recursive sets.
    (c) The class of the recursive sets in $P/poly$.
    (d) The class of the recursive NP-hard sets.

4. Use the Uniform Diagonalization Theorem to show that the following classes are *not* recursively presentable, unless they are empty:

    (a) $NP - P$.
    (b) The non-complete sets in NP.
    (c) The non-complete sets in $NP - P$.
    (d) The analogous classes for $PSPACE$ instead of NP.
    (e) The class of all the infinite sets in P.

5. Let $C_1, C_2, \ldots$ be a sequence of recursively presentable classes. Denote by $f_1, f_2, \ldots$ the recursive functions that enumerate the corresponding presentations, and assume that a machine computing f_i can be found effectively from i. Show that the union of all the classes C_i is recursively presentable.

6. Let C_1 and C_2 be two recursively presentable classes. Assume that their intersection, C_3, contains some set B together with all the finite variations of B. Show that C_3 is recursively presentable.

7. Let C_1 be a class closed under join and under polynomial time m-reducibility. Assume that C_2 and C_3 are recursively presentable classes, closed under finite variations, and that $C_1 = C_2 \cup C_3$. Show that either $C_1 = C_2$ or $C_1 = C_3$.

8. Let C_1 and C_2 be two nonempty recursively presentable classes closed under finite variations. Assume that C_1 is closed under join and under polynomial time m-reducibility. Show that $C_1 - C_2$ is not recursively presentable, unless it is empty.

9. For any set A define the class $(\leq A)$ as the class of all sets m-reducible to A in polynomial time. Show that the following statement is equivalent to the Uniform Diagonalization Theorem: for any two recursively presentable classes C_1 and C_2 closed under finite variations, if $(\leq A) \not\subseteq C_1$ and $(\leq A) \not\subseteq C_2$, then $(\leq A) \not\subseteq C_1 \cup C_2$.

10. Let C_1 and C_2 be two recursively presentable classes. Assume that C_1 contains only infinite sets and that C_2 is closed under finite variations. Let B be any set not in C_2. Show that there is a set D in P such that $B \cap D$ is neither in C_1 nor in C_2.

11. Most of the corollaries to the Uniform Diagonalization Theorem stated in this chapter can be obtained from the previous exercise. Find them and prove them in this way.

12. Let C_1 be a recursively presentable class containing only infinite sets. Let C_2 be any class closed under polynomial time m-reducibility, which contains some infinite set B different from Γ^*. Show that the subclass of the infinite sets in C_2 is not included in C_1 (no matter how big C_1 is). (Hint: use Exercises 1 and 10.)

13. Show that if a class C is recursively presentable, is closed under intersection, and contains P, then C has no maximal set under the partial ordering of inclusion modulo finite variations.

14. Obtain analogues of Theorem 7.10 and Corollary 7.11 for the m-reducibility.

15. Show: if C is recursively presentable, then its closure under finite variants is also recursively presentable.

7.6 Bibliographical Remarks

The existence of sets in NP which are neither in P nor NP-complete, under the hypothesis $P \neq NP$, was shown in Ladner (1975a). This work also contains the construction of two incomparable sets in NP, and the construction of minimal pairs for the m-reducibility. Later, in Balcázar and Díaz (1982), the existence of infinite families of incomparable sets in NP was shown. Both references prove their results by direct, quite involved constructions.

The definition of the set $G[r]$ and its use for obtaining results similar to Ladner's were presented in Landweber, Lipton, and Robertson (1981). Based on this work, the Uniform Diagonalization Theorem was stated in Schöning (1982). We have followed mainly this last reference. Similar, but a little more obscure theorems appeared in Chew, Machtey (1981); we have presented the most important of them in Exercise 10.

Very strong generalizations of these arguments, which can be used for working with classes below P, are presented in Schmidt (1985). The analysis made in this reference shows exactly where this kind of theorem holds and where it does not. Two particular cases of the results of this last reference are Exercises 8 and 9.

Exercise 13 is an example of an application of the Uniform Diagonalization Theorem to the lattice arising from the inclusion modulo finite variants. It was proved for NP in Homer (1981).

We should mention that similar theorems for other reducibilities (in particular, length-increasing invertible m-reductions) have been found in Regan

(1983), and that a theorem of similar flavor for obtaining minimal pairs for P is presented in Schöning (1984a). Finally, let us indicate that the technique used has been shown to have some limitations. A discussion on these limitations, based on the interesting concept of "inhomogeneity", can be found in Ambos-Spies (1986).

8 The Polynomial Time Hierarchy

8.1 Introduction

In the previous chapter, we have seen that under the hypothesis $P \neq NP$ there are "hierarchies" of incomparable sets between P and the NP-complete sets. In this section we are going to study a very different kind of hierarchy between P and $PSPACE$: the polynomial time hierarchy.

The polynomial time hierarchy is the polynomial version of the Kleene hierarchy studied in Recursive Function Theory. We begin the chapter by introducing the definitions and basic properties. There follows a section in which the polynomial time hierarchy is characterized by means of alternating quantifiers, in a very similar manner to the Kleene hierarchy. In the next section we study the concept of completeness for each of the different levels of the hierarchy, showing the existence of complete sets in every class. The last section of the chapter deals with the relation between the probabilistic class BPP, defined in Chapter 6, and the polynomial time hierarchy.

The polynomial time hierarchy will play an important role in many of the chapters in Volume II.

8.2 Definition and Properties

Definition 8.1 *The polynomial time hierarchy is the structure formed by the classes Σ_k, Π_k, and Δ_k for each $k \geq 0$, where*

1. $\Sigma_0 = \Pi_0 = \Delta_0 = P$;
2. $\Sigma_{k+1} = NP(\Sigma_k)$ for $k \geq 0$;
3. $\Delta_{k+1} = P(\Sigma_k)$ for $k \geq 0$;
4. $\Pi_{k+1} = co\text{-}NP(\Sigma_k)$ for $k \geq 0$.

Define also $PH = \bigcup_{k \geq 0} \Sigma_k = \bigcup_{k \geq 0} \Pi_k = \bigcup_{k \geq 0} \Delta_k$

It is usual to provide the classes Σ, Π, and Δ of the polynomial time hierarchy with a P superscript in order to distinguish them from the analogously denoted classes of the Kleene (arithmetic) hierarchy. However, the arithmetic hierarchy will not be used in this book. Hence no confusion may arise here, and we will drop the superscript.

We define also the relativized version of the polynomial time hierarchy as follows.

Definition 8.2 *The polynomial time hierarchy relative to an oracle A is the structure formed by the classes $\Sigma_k(A)$, $\Pi_k(A)$, and $\Delta_k(A)$ for each $k \geq 0$, where*

1. $\Sigma_0(A) = \Pi_0(A) = \Delta_0(A) = P(A)$;
2. $\Sigma_{k+1}(A) = NP(\Sigma_k(A))$ for $k \geq 0$;
3. $\Delta_{k+1}(A) = P(\Sigma_k(A))$ for $k \geq 0$;
4. $\Pi_{k+1}(A) = co\text{-}NP(\Sigma_k(A))$ for $k \geq 0$.

Define also $PH(A) = \bigcup_{k \geq 0} \Sigma_k(A) = \bigcup_{k \geq 0} \Pi_k(A) = \bigcup_{k \geq 0} \Delta_k(A)$

We study in this chapter the unrelativized version of the polynomial time hierarchy. However, most of the results we are going to present hold for the relativized versions. See the exercises at the end of this chapter.

We start with some properties of the polynomial time hierarchy. The proof of the following proposition uses many results established in Chapter 1.

Proposition 8.3
(a) $\Delta_1 = P$
(b) $\Pi_k = co\text{-}\Sigma_k$ for $k \geq 0$
(c) $\Sigma_{k+1} = NP(\Pi_k)$ for $k \geq 0$
(d) $\Delta_{k+1} = P(\Pi_k)$ for $k \geq 0$
(e) $\Pi_{k+1} = co\text{-}NP(\Pi_k)$ for $k \geq 0$
(f) $\Sigma_{k+1} = NP(\Delta_{k+1})$ for $k \geq 0$
(g) $\Pi_{k+1} = co\text{-}NP(\Delta_{k+1})$ for $k \geq 0$

Proof.

(a) By definition, $\Delta_1 = P(P)$. Therefore, the result follows from Proposition 4.11.
(b) Immediate from the definition.
(c) Exchange in each machine the state "continue after query with answer YES" with the state "continue after query with answer NO". Then use as oracle the complement of the old oracle.
(d) Same as (c).
(e) Same as (c).
(f) The inclusion from left to right is immediately clear. For the converse, assume $B \in \Delta_{k+1}$. Let M_0 be a deterministic machine accepting B with the aid of some oracle in Σ_k. From each machine M_1 using B as oracle, construct a new machine M_2 which simulates M_1, calling M_0 as a subroutine every time a query is made. Now M_2 accepts the same set as M_1, but its oracle is in Σ_k instead of Δ_{k+1}.
(g) Take complements in (f). □

Proposition 8.4   *For every $k \geq 0$,*

(a) Δ_k *is closed under complementation.*

(b) $P(\Delta_k) = \Delta_k$, *so that Δ_k is closed under polynomial time T-reducibility.*

(c) $\Sigma_k \bigcup \Pi_k \subseteq \Delta_{k+1}$.

(d) $\Delta_k \subseteq \Sigma_k \bigcap \Pi_k$.

(e)   *All the classes in the polynomial time hierarchy are closed under polynomial time m-reducibility.*

(f)   *All the classes in the polynomial time hierarchy are closed under the "join" operator.*

(g)   *All the classes in the polynomial time hierarchy are closed under union and intersection.*

(h)   *If either $\Sigma_k \subseteq \Pi_k$ or $\Pi_k \subseteq \Sigma_k$ then $\Sigma_k = \Pi_k$.*

Proof.

(a)   Exchange accepting and rejecting states in the deterministic oracle machines defining Δ_k.

(b)   For $k = 0$, $P(\Delta_k) = P(P)$. Substituting polynomial time computations for the oracle queries, convert the polynomial time machines with oracle into non-oracle machines. The general case is analogous, observing that $P(\Delta_k) = P(P(\Sigma_{k-1}))$.

(c)   Consider oracle machines that query the input to the oracle, and accept according to the answer (respectively, contrarily to the answer).

(d)   Follows from the obvious fact that for every A, $A \in NP(A)$ and $A \in$ co-$NP(A)$.

(e)   Follows from the fact that for every A the classes $P(A)$ and $NP(A)$ are both closed under the m-reducibility.

(f)   Same reason as (e).

(g)   Same reason as (e).

(h)   By Lemma 1.15 and Proposition 8.3(b).   □

The last proposition shows that PH has the following inclusion structure: for all $k \geq 0$,

$$\Sigma_k \bigcup \Pi_k \subseteq \Delta_{k+1} \subseteq \Sigma_{k+1} \bigcap \Pi_{k+1}$$

It is not known whether any of these inclusions is proper. However, it will be shown later that if any of these is an equality then so are all the inclusions at upper levels.

We close this section locating the polynomial time hierarchy among the complexity classes we have studied in previous chapters.

Theorem 8.5 $PH \subseteq PSPACE$.

Proof. We prove by induction on k that, for each k, $\Sigma_k \subseteq PSPACE$. The inclusion of PH follows.

Case $k = 0$. Σ_0 is the class P by definition. We already know that P is included in $PSPACE$.

From k to $k + 1$. Assume that Σ_k is included in $PSPACE$. As Σ_{k+1} is defined as $NP(\Sigma_k)$, we deduce that it is contained in $NP(PSPACE)$. But $NP(PSPACE) \subseteq PSPACE(PSPACE) = PSPACE$, where the inclusion follows from the fact that $NP(A)$ is always contained in $PSPACE(A)$, as shown in Proposition 4.11(g). $\qquad\qquad\square$

8.3 Characterization and Consequences

Let us give a characterization of the polynomial time hierarchy in terms of alternating bounded quantifiers. From this characterization we shall deduce more properties of the PH. We introduce the following notation: for any property $R(x)$ on strings x, and some polynomial $p(\cdot)$,

(a) $\exists^{p(n)} x R(x)$ means that there is a x such that $|x| \leq p(n)$ having the property R. We will omit n if it is clear from the context, and just write $\exists^p x R(x)$.

(b) $\forall^{p(n)} x R(x)$ means that every x with $|x| \leq p(n)$ has the property R. Again we just write $\forall^p x R(x)$ when n is clear from the context.

(c) For any class $\mathcal{C}$ of sets, the class $\exists \mathcal{C}$ is the one formed by all the sets A which can be defined by: $x \in A$ if and only if $\exists^{p(|x|)} y \langle x, y \rangle \in B$, where B is some element of $\mathcal{C}$ depending only on A.

(d) In an analogous way, $\forall \mathcal{C}$ is the class formed for all the sets A which can be defined by: $x \in A$ if and only if $\forall^{p(|x|)} y \langle x, y \rangle \in B$, where B is some element of $\mathcal{C}$ depending only on A.

(e) Let A be any set. We denote by $A^{(*)}$ the following set:

$$A^{(*)} = \{ x \mid x = \langle y_1, y_2, \ldots, y_n \rangle \text{ for some } n, \text{ and for each } i \leq n, y_i \in A \}$$

For the next proposition, we say that $\mathcal{C}$ is closed under pairing if and only if for every set $A \in \mathcal{C}$, the set of all the words $\langle x, y \rangle$ where $x \in A$ is also in $\mathcal{C}$. All our classes are closed under pairing. (Compare with the concept of cylindrification in Volume II.)

Proposition 8.6

(a) *For every class $\mathcal{C}$, $A \in \exists \mathcal{C}$ if and only if $\overline{A} \in \forall co\text{-}\mathcal{C}$, i.e., $co\text{-}\exists \mathcal{C}$ is exactly $\forall co\text{-}\mathcal{C}$.*

(b) *For every class $\mathcal{C}$ which is closed under pairing, $\mathcal{C} \subseteq \forall \mathcal{C}$ and $\mathcal{C} \subseteq \exists \mathcal{C}$.*

Proof.

(a) Assuming that $x \in A$ if and only if $\exists^{p(|x|)}y\langle x, y\rangle \in B$, and negating both sides of the equivalence, we get $x \in \overline{A}$ if and only if $\neg\exists^{p(|x|)}y\langle x, y\rangle \in B$, if and only if $\forall^{p(|x|)}y\neg\langle x, y\rangle \in B$, if and only if $\forall^{p(|x|)}y\langle x, y\rangle \in \overline{B}$.

(b) Just add dummy quantifiers.   □

The next theorem indicates that the power of the bounded existential quantifier is equivalent to the polynomial time nondeterministic Turing machines, and will be used below in the "alternating quantifiers" characterization of the polynomial time hierarchy. It provides a characterization of the classes obtained by quantification of the classes in the polynomial time hierarchy. First we prove a technical lemma.

Lemma 8.7 *Let C be any class Σ_k, Π_k, or Δ_k of the polynomial time hierarchy. Then for every A, $A \in C$ if and only if $A^{(*)} \in C$.*

Proof. We can m-reduce A to $A^{(*)}$ via the function mapping each y to $\langle y\rangle$. Hence, if $A^{(*)}$ is in C then A is in C by the closure under the m-reducibility.

Now assume that A is in C, where C is a class Δ_k. Consider a deterministic polynomial time machine M, which for the case of $k \geq 1$ is allowed to query an oracle in Σ_{k-1}, and has no oracle access if $k = 0$. The algorithm of Figure 8.1 witnesses that $A^{(*)}$ is in C.

```
input x = ⟨y₁,...,yₙ⟩
for i := 1 to n do
      simulate M on yᵢ
      if M rejects then reject
      end for
accept
end
```

Figure 8.1 An algorithm for $A^{(*)}$

Observe that the case of Δ_0 proves the lemma also for Σ_0 and Π_0. The case of Σ_k, $k \geq 1$, uses the same algorithm, which is nondeterministic because M will be nondeterministic.

Finally, for the case of Π_k, $k \geq 1$, let M be a nondeterministic machine for $\overline{A}$ with an oracle in Σ_{k-1}. Consider the algorithm of Figure 8.2.

It accepts the complement of $A^{(*)}$, and shows that it is in Σ_k. Hence, $A^{(*)}$ belongs to Π_k.   □

We move now to the theorem.

```
input x = ⟨y₁, ..., yₙ⟩
for i := 1 to n do
      simulate M on yᵢ
      if M accepts then accept
      end for
reject
end
```

Figure 8.2   An algorithm for the complement of $A^{(*)}$

Theorem 8.8
(a) $\exists P = NP$
(b) $\forall P = co\text{-}NP$
(c) $\exists \Sigma_k = \Sigma_k$ for $k > 0$
(d) $\forall \Pi_k = \Pi_k$ for $k > 0$
(e) $\exists \Pi_k = \Sigma_{k+1}$ for $k \geq 0$
(f) $\forall \Sigma_k = \Pi_{k+1}$ for $k \geq 0$

Proof.

(a) Let A be a set in $\exists P$. Then there is a set B in P and a polynomial p such that $x \in A$ if and only if $\exists^{p(|x|)} y \langle x, y \rangle \in B$. Consider a nondeterministic machine operating as follows:

```
read x
guess y with |y| ≤ p(|x|)
if ⟨x, y⟩ ∈ B then accept
end
```

The test $\langle x, y \rangle \in B$ can be made in polynomial time; hence this machine witnesses the fact that $A \in NP$. Conversely, let A be a set in NP, and let M be a nondeterministic machine accepting A in polynomial time. Let p be a polynomial large enough to write down with $p(n)$ symbols every computation of M on inputs of length n. Then $x \in A$ if and only if $\exists^{p(|x|)} y$ ("y encodes an accepting computation of M on x"). The predicate in parentheses is decidable in polynomial time. This shows that A is in $\exists P$.

(b) By part (a) and Proposition 8.6(a).
(c) By Proposition 8.6(b), $\Sigma_k \subseteq \exists \Sigma_k$. Let us show the converse. Let $A \in \exists \Sigma_k$. By definition, there is a $B \in \Sigma_k$ such that $x \in A$ if and only if $\exists^{p(|x|)} y \langle x, y \rangle \in B$ for some polynomial p. By the definition of Σ_k, we have that $B \in NP(\Sigma_{k-1})$. Therefore there exists a nondeterministic oracle Turing machine M_1 and an oracle D in Σ_{k-1} such that $B =$

$L(M_1, D)$. Define a nondeterministic oracle Turing machine M_2 which operates as follows:

read x
guess y with $|y| \leq p(|x|)$
accept x if and only if M_1 accepts $\langle x, y \rangle$

With oracle D, M_2 accepts A in polynomial time, and therefore we have that $A \in NP(D) \subseteq NP(\Sigma_{k-1}) = \Sigma_k$.

(d) By part (c) and Proposition 8.6(a).
(e) Let A be a set in $\exists \Pi_k$. Then for some $B \in \Pi_k$, $x \in A$ if and only if $\exists^{p(|x|)} y \langle x, y \rangle \in B$. Consider a nondeterministic oracle machine M that on input x just guesses y of length at most $p(|x|)$ and then queries $\langle x, y \rangle$ to its oracle, accepting if and only if the answer is YES. This machine accepts A in polynomial time with oracle B. Hence A is in $NP(\Pi_k)$, which is Σ_{k+1}. Thus $\exists \Pi_k \subseteq \Sigma_{k+1}$.
Conversely, we prove inductively that $\Sigma_{k+1} \subseteq \exists \Pi_k$.

Case $k = 0$. Follows from part (a).

From k to $k+1$. Assume that it is true for Σ_k, $k \geq 0$, and let $A \in \Sigma_{k+1}$. Then $A = L(M, B)$ for some $B \in \Sigma_k$ and some nondeterministic oracle Turing machine M working in polynomial time. Therefore $x \in A$ if and only if there is a computation of M with oracle B which accepts x. Let p be a polynomial large enough to write down with $p(n)$ symbols every computation of M on inputs of length n. Then

$\quad$ $x \in A$ if and only if
$\qquad$ $\exists^{p(|x|)} y$ $\quad$ ($\quad$ [y encodes an accepting computation of M on x
$\qquad\qquad\qquad\qquad$ with queries $z_1, z_2, \ldots, z_i$ answered YES
$\qquad\qquad\qquad\qquad$ and with queries $w_1, w_2, \ldots, w_j$ answered NO]
$\qquad\qquad\qquad\quad$ and $\langle z_1, z_2, \ldots, z_i \rangle \in B^{(*)}$
$\qquad\qquad\qquad\quad$ and $\langle w_1, w_2, \ldots, w_j \rangle \in \overline{B}^{(*)}$)

(Note that, in order to be completely formal, the z's and the w's should be existentially quantified at the same level as y.) The predicate in square brackets is decidable in polynomial time; thus it is in Π_k. As B is in Σ_k, by Lemma 8.7, $B^{(*)}$ is in Σ_k, $\overline{B}$ is in Π_k, and $\overline{B}^{(*)}$ is in Π_k. By the induction hypothesis, membership to $B^{(*)}$ can be expressed as: $u \in B^{(*)}$ if and only if $\exists^p t \langle u, t \rangle \in D$, where $D \in \Pi_{k-1}$, and hence is also in Π_k. Thus

$\quad$ $x \in A$ if and only if
$\qquad$ $\exists^{p(|x|)} v$ $\quad$ ($\quad$ [$v = \langle y, t \rangle$ and
$\qquad\qquad\qquad\qquad$ y encodes an accepting computation of M on x

with queries $z_1, z_2, \ldots, z_i$ answered YES
and with queries $w_1, w_2, \ldots, w_j$ answered NO]
and $\langle\langle z_1, z_2, \ldots, z_i \rangle, t\rangle \in D$
and $\langle w_1, w_2, \ldots, w_j \rangle \in \overline{B}^{(*)}$)

The predicate in parentheses is now the conjunction of three predicates in Π_k, hence is also in Π_k. Therefore A has been expressed as existential quantification of Π_k: $A \in \exists\Pi_k$.

(f) Just take complements in (e). □

We are ready for the main characterization of the polynomial time hierarchy.

Theorem 8.9 *For any $k \geq 0$,*

(a) *$A \in \Sigma_k$ if and only if there is a set B in P and a polynomial p such that*

$$x \in A \text{ if and only if } \exists^{p(|x|)}y_1 \forall^{p(|x|)}y_2 \ldots Qy_k \langle x, y_1, \ldots, y_k \rangle \in B$$

where Q is $\forall^{p(|x|)}$ if k is even, while Q is $\exists^{p(|x|)}$ if k is odd.

(b) *$A \in \Pi_k$ if and only if there is a set B in P and a polynomial p such that*

$$x \in A \text{ if and only if } \forall^{p(|x|)}y_1 \exists^{p(|x|)}y_2 \ldots Qy_k \langle x, y_1, \ldots, y_k \rangle \in B$$

Here the quantifiers alternate between existential and universal as in (a).

Observe that statements (a) and (b) in Theorem 8.8 are precisely the particularization of this theorem to the case $k = 1$.

Proof. The proof is by induction on k, and proving simultaneously both statements (a) and (b) for each k.

Case $k = 0$. By definition, $\Sigma_0 = \Pi_0 = P$. This proves (a) and (b) for this case, since the number of alternating quantifiers is zero.

From $k - 1$ to k. We prove (a); then (b) follows by complementation using Proposition 8.3(b). Let A be in Σ_k. By the previous theorem, A is in $\exists\Pi_{k-1}$. By definition, there is a B in Π_{k-1} such that $x \in A$ if and only if $\exists^p y \langle x, y \rangle \in B$. By the induction hypothesis, the fact that $\langle x, y \rangle \in B$ can be expressed as: $\langle x, y \rangle \in B$ if and only if $\forall^p z_1 \exists^p z_2 \ldots \langle\langle x, y \rangle, z_1, \ldots, z_{k-1}\rangle \in D$, where D is some set in P. Without loss of generality we assume that the polynomial is the same. Let D' be a set such that $\langle x, y, z_1, \ldots, z_{k-1}\rangle \in D'$ if and only if $\langle\langle x, y \rangle, z_1, \ldots, z_{k-1}\rangle \in D$.

As D is in P, D' is also in P. Then we can write: $x \in A$ if and only if $\exists^p y \langle x, y \rangle \in B$, if and only if $\exists^p y \forall^p z_1 \exists^p z_2 \ldots \langle x, y, z_1, \ldots, z_{k-1} \rangle \in D'$, which is the required form for the definition of A.   □

From the above characterizations, we can easily obtain some important results about the polynomial time hierarchy and its relationship to other complexity classes.

As we have already mentioned, it is an open problem to show any proper inclusion between successive classes of the polynomial time hierarchy. However, as a corollary of the next result, we shall show that proper inclusion between some pair of successive classes implies $P \neq NP$.

Theorem 8.10   If $\Sigma_k = \Pi_k$ for some $k \geq 1$, then $\Sigma_{k+j} = \Pi_{k+j} = \Sigma_k$ for all $j \geq 0$.

Proof.   We prove it by induction on j.

Case $j = 0$. The result is just what the hypothesis states.

From j to $j+1$. Assume that $\Sigma_{k+j} = \Pi_{k+j} = \Sigma_k$. Then, by Theorem 8.8, parts (e) and (c), $\Sigma_{k+j+1} = \exists \Pi_{k+j} = \exists \Sigma_{k+j} = \Sigma_{k+j}$, which is Σ_k by the induction hypothesis. Furthermore, taking complements in the equalities, $\Pi_{k+j+1} = \Pi_{k+j}$, which is also Σ_k by the induction hypothesis. Thus, $\Sigma_{k+j+1} = \Pi_{k+j+1} = \Sigma_k$.   □

Corollary 8.11   If $P = \Sigma_0 \neq \Sigma_k$ for some $k \geq 1$, then $P \neq NP$.

Corollary 8.12   Either for all $k \geq 0$, $\Sigma_k \neq \Sigma_{k+1}$, or the polynomial hierarchy consists only of finitely many different classes Σ_k.

The second case in the last corollary is called the "collapse" of the polynomial time hierarchy. If the hierarchy is proper up to level $i - 1$, but $\Sigma_i = \Pi_i$, the hierarchy is said to "collapse at level i".

The following theorem proves a collapse of the polynomial time hierarchy if it is the case that $PH = PSPACE$.

Theorem 8.13   If $PH = PSPACE$ then $\Sigma_k = \Sigma_{k+1}$ for some k.

Proof.   Recall that $PSPACE$ has complete sets with respect to polynomial time m-reducibility, as for example QBF in Theorem 3.29. If $PH = PSPACE$ then QBF $\in PH$ and by the definition of PH, QBF $\in \Sigma_k$ for some k, which implies that every set in $PSPACE$ is m-reducible to a set in Σ_k. In particular, all the sets in Π_k are reducible to this set. But Σ_k is closed under polynomial time m-reducibility, hence $\Pi_k \subseteq \Sigma_k$. By Proposition 8.4(h), this implies that $\Pi_k = \Sigma_k$, and the hierarchy collapses.   □

8.4 Complete Sets and Presentability

As a consequence of Theorem 4.16, all the classes in the polynomial hierarchy have complete sets. This is because each of them can be expressed as $P(A)$ or $NP(A)$ or co-$NP(A)$ where the set A is the complete set for the level immediately below. We state these properties more formally in our next proposition. See the exercises for other complete sets in the polynomial time hierarchy.

Proposition 8.14 *For any $k > 0$, if A is Σ_k-complete, then $K(A)$ is Σ_{k+1}-complete.*

Proof. As A is in Σ_k and $K(A)$ is always in $NP(A)$, we have that $K(A)$ is in Σ_{k+1}. As A is complete for Σ_k, every set in Σ_k is m-reducible to A. Let B be a set in $\Sigma_{k+1} = NP(\Sigma_k)$. There is a set D in Σ_k such that $B \in NP(D)$. But D is m-reducible to A, so that $B \in NP(A)$. By the completeness of $K(A)$ in $NP(A)$ it follows that B is m-reducible to $K(A)$. Therefore $K(A)$ is complete in Σ_{k+1}. □

Following the standard use, define $K^1(A)$ as $K(A)$, $K^n(A)$ as $K(K^{n-1}(A))$, and K^n as $K^n(\emptyset)$.

Corollary 8.15 *For $n > 0$, K^n is Σ_n-complete.*

We turn to the recursive presentability of the classes in the polynomial time hierarchy, as defined in Chapter 7.

Lemma 8.16 *All the classes in the polynomial time hierarchy are recursively presentable.*

Proof. For all n, the class Δ_n is $P(K^n)$ by the completeness of this set. By Lemma 7.15, it is recursively presentable. Also Σ_0 is recursively presentable, because it is P by definition.

For all $n \geq 1$, the class Σ_n is $NP(K^{n-1})$. Thus, it is recursively presentable by Proposition 7.8(h).

For all $n \geq 0$, the class Π_n is the class of complements of Σ_n. Thus, it is recursively presentable by Lemma 7.6.

Observe that the recursive presentations of the classes Σ_k obtained from the results of Chapter 7 can be found from k in an effective way. Thus, the presentability of PH follows from Exercise 5 in Chapter 7. □

Recall from Proposition 7.8(g) that the class of $PSPACE$-complete sets is recursively presentable. Once the presentability of these classes has been shown, we can apply the Uniform Diagonalization Theorem to obtain the following results.

Proposition 8.17 *For any $k \geq 0$, if $PSPACE \neq \Sigma_k$ then there exist sets in PSPACE which are not PSPACE-complete and which are not in Σ_k.*

Proof. If $PSPACE \neq \Sigma_k$ then QBF $\notin \Sigma_k$. Hence, taking $A_1 = \emptyset$, $A_2 =$ QBF, $\mathcal{C}_1$ the class of *PSPACE*-complete sets, and $\mathcal{C}_2 = \Sigma_k$, by the Uniform Diagonalization Theorem there exists a set A which is neither in Σ_k, nor *PSPACE*-complete, but $A \leq_m$ QBF, so that $A \in PSPACE$.   □

Other applications of the Uniform Diagonalization Theorem to the polynomial time hierarchy are shown in Exercises 3 to 8.

8.5 *BPP* and the Polynomial Time Hierarchy

In this section we present the inclusion of the probabilistic class *BPP* in the polynomial time hierarchy. More precisely, we show that it is included in the second level of the hierarchy.

Recall that the class *BPP* has been defined in Chapter 6 as the class of sets that can be decided in polynomial time by probabilistic machines with bounded error probability, and that this class has been characterized in Theorem 6.22. We will use this characterization here.

Before proceeding to the proof, we need some quite technical auxiliary definitions and properties.

Definition 8.18 *Given a polynomial p and a set E, we say that E is p-biased if and only if for each n and for each y of length $p(n)$, either*

(a) *there are at most $2^{p(n)-n}$ words w of length $p(n)$ such that $\langle y, w \rangle \in E$, or*

(b) *there are at most $2^{p(n)-n}$ words w of length $p(n)$ such that $\langle y, w \rangle \notin E$.*

This definition formalizes a property that we will use later, which can be stated intuitively as follows: for each y of length $p(n)$, either E contains "many" words $\langle y, w \rangle$ with $w \in \Gamma^{p(n)}$, or "only a few" of them. The interest in this kind of sets is that, for each y, we can find out whether (a) or (b) above holds, by means of a set which is not much more difficult than E itself. More precisely, this "discriminating" set is in $\Sigma_2(E)$. It is defined as follows.

Definition 8.19 *Given a polynomial p and a set E, define the following set:*

$$most(E, p) = \{ y \mid |y| = p(n), \text{ and there are more than } 2^{p(n)-n} \text{ words } w$$

$$\text{of length } p(n) \text{ such that } \langle y, w \rangle \in E \}$$

The indicated property of this set $most(p, E)$ is presented in the following lemma, whose proof presents a nontrivial way of producing sets which lie in relativizations of Σ_2.

Lemma 8.20 *For every polynomial p and any p-biased set E, the set most (E, p) is in $\Sigma_2(E)$. Moreover, if E is in Π_1 then $most(E, p)$ is in Σ_2.*

Proof. The proof involves two counting arguments. Throughout this proof, we denote by $u \oplus v$ the bitwise addition of u and v, where both u and v belong to $\{0, 1\}^n$, i.e., the ith bit of $u \oplus v$ is the exclusive "OR" of the ith bits of u and v. For a set $B \subseteq \{0, 1\}^n$, and a word u of $\{0, 1\}^n$, the u-*translation* of B is the set

$$\{v \mid \exists w \in B \text{ such that } v = w \oplus u\}$$

i.e., the set of words of $\{0, 1\}^n$ that can be reached from B by bitwise addition of u to all the words of B. Notice that all the translations of B have the same cardinality, namely $|B|$.

The idea is that B is able to cover all of $\{0, 1\}^n$ by only polynomially many translations if and only if it has "a lot of" words. Let us proceed to formalize this argument. Observe that $v = w \oplus u$ if and only if $w = v \oplus u$. Therefore, the u-translation of B can be defined equivalently as

$$\{v \mid v \oplus u \in B\}$$

which has no quantifiers. Using this fact, we will prove the following claim.
Claim. For sufficiently large n, the following holds: $y \in most(E, p)$ with $|y| = p(n)$ if and only if

$$\exists u = \langle u_1, \ldots, u_{p(n)} \rangle, |u_i| = p(n), \forall v, |v| = p(n), (\bigvee_{i=1}^{p(n)} \langle y, v \oplus u_i \rangle \in E)$$

Before proving it, observe that the meaning of the right hand side is the following: taking the set

$$B(y) = \{w \mid |w| = |y| = p(n), \langle y, w \rangle \in E\}$$

it says that there are polynomially many u-translations of $B(y)$ such that their union contains every word of $\{0, 1\}^{p(n)}$. As we have already said, this will be equivalent to saying that $B(y)$ contains "many" words, which is $most(E, p)$ by definition.

We prove now this equivalence. Assume that

$$\exists u = \langle u_1, \ldots, u_{p(n)} \rangle, |u_i| = p(n), \forall v, |v| = p(n), (\bigvee_{i=1}^{p(n)} \langle y, v \oplus u_i \rangle \in E)$$

Note that $\langle y, v \oplus u_i \rangle \in E$ if and only if $v \oplus u_i \in B(y)$, which holds if and only if v belongs to the u_i-translation of $B(y)$. Thus, we have that every word v of $\Gamma^{p(n)}$ belongs to some of these u_i-translations. As there are $2^{p(n)}$

such words v, but just $p(n)$ u_i-translations, each of cardinality $|B(y)|$, we obtain that

$$|B(y)| \geq \frac{2^{p(n)}}{p(n)} > \frac{2^{p(n)}}{2^n} = 2^{p(n)-n}$$

Thus, by the definition of $B(y)$, there are more than $2^{p(n)-n}$ words w in $\Gamma^{p(n)}$ with $\langle y, w \rangle \in E$, which is the definition of $y \in most(E, p)$. Conversely, assume now that

$$\forall u = \langle u_1, \ldots, u_{p(n)} \rangle, |u_i| = p(n), \exists v, |v| = p(n), (\bigwedge_{i=1}^{p(n)} \langle y, v \oplus u_i \rangle \notin E)$$

For each word v of $\Gamma^{p(n)}$, define the set

$$\text{fooled-by}(v) = \{u = \langle u_1, \ldots, u_{p(n)} \rangle \mid \forall i, \langle y, v \oplus u_i \rangle \notin E\}$$

We have that every u is in the set fooled-by(v) for some v. Note that there are $2^{p^2(n)}$ ways of choosing u, and hence, for some v, $|\text{fooled-by}(v)| \geq 2^{p^2(n)-p(n)}$. Fix this v, and let F be the set of all $w \in \Gamma^{p(n)}$ such that $\langle y, w \rangle \notin E$. Each u in fooled-by(v) is of the form $\langle u_1, \ldots, u_{p(n)} \rangle$ where $\forall i \langle y, v \oplus u_i \rangle \notin E$, and therefore $\langle y, v \oplus u_i \rangle \in F$. Hence, u can be constructed in at most $|F|^{p(n)}$ ways. This implies that

$$|F|^{p(n)} \geq |\text{fooled-by}(v)| \geq 2^{p^2(n)-p(n)}$$

and therefore $|F| \geq 2^{p(n)-1} > 2^{p(n)-n}$.

Now we have that there are more than $2^{p(n)-n}$ words $w \in \Gamma^{p(n)}$ such that $\langle y, w \rangle \notin E$. But E is p-biased, and we have just proved that condition (b) in the definition of p-biased fails for this y. Hence, condition (a) must hold, and by definition of $most(E, p)$ we obtain that $y \notin most(E, p)$. This completes the proof of the claim.

With this claim we have obtained a definition of $most(E, p)$ by means of a $\Sigma_2(E)$ predicate. Now assume that $E \in \Pi_1$. Then a Π_1 predicate can be substituted for the test for membership of E in the statement of the claim, and the universal quantifier of this Π_1 predicate can be merged with the "$\forall v$" to obtain an unrelativized predicate Σ_2. Thus, in this case, $most(E, p) \in \Sigma_2$. This completes the proof of the lemma.   □

Now we are ready for the main theorem of this section: the inclusion of BPP in the second level of the polynomial time hierarchy.

Theorem 8.21 $BPP \subseteq \Sigma_2 \cap \Pi_2$.

Proof.   We show that BPP is included in Σ_2. It follows that co-BPP is included in co-Σ_2. But $BPP =$ co-BPP by the closure under complementation of this class (Proposition 6.19), and co-$\Sigma_2 = \Pi_2$ by definition. Thus, *BPP* is also included in Π_2.

Let A be a set in *BPP*. Apply Theorem 6.22 with $q(n) = n$, and obtain a set B in P and a polynomial p such that for every n, there are at least $2^{p(n)} \cdot (1 - (\frac{1}{2})^n)$ words y of length $p(n)$ such that $x \in A$ if and only if $\langle x, y \rangle \in B$ for all x of length n.

Using this B and polynomial p, we can informally express membership of A as follows:

$x \in A$ if and only if $\exists^{p(|x|)} y (\text{``}y \text{ is a correct advice''}$ and $\langle x, y \rangle \in B)$

We need to express the quoted predicate as a Σ_2 predicate. We do this as follows: we know that there are "many" (at least $2^{p(n)} \cdot (1 - (1/2)^n)$) words of length $p(n)$ which are correct advices. Thus, if we find more than $2^{p(n)-n}$ such words w that "agree" on the set accepted by B with advice w, as not all of them are incorrect (because there are only $2^{p(n)-n}$ incorrect ones), all of them must be correct. Then, we just have to check that y "agrees" with all of them.

Formally, y is a correct advice for A at length n if and only if there are more than $2^{p(n)-n}$ words w of length $p(n)$ which "agree" with y on the set accepted by B when given as advice. Thus membership of A can be expressed as follows: $x \in A$ if and only if $\exists^{p(|x|)} y \langle x, y \rangle \in B$ and there are more than $2^{p(|x|)-|x|}$ words w of length $p(|x|)$ such that

$$\forall z, |z| = |x|, \langle z, y \rangle \in B \text{ if and only if } \langle z, w \rangle \in B$$

Note that because of the way in which p and B are chosen, there is always a y which "agrees" with more than $2^{p(n)-n}$ words w. Furthermore, any such y is a correct advice for A at length n. In order to transform this expression into a Σ_2 predicate, we use our previous definition and lemma, taking as the set E the following set:

$$E = \{\langle y, w \rangle \mid |y| = |w| = p(n) \text{ and } \forall z |z| = n,$$
$$\langle z, y \rangle \in B \text{ if and only if } \langle z, w \rangle \in B\}$$

Note that E is a p-biased set, because if y is a correct advice then all the correct w agree with it, while if y is not correct then only some incorrect ones may agree with it. Note also that E is in Π_1, as it is defined by polynomially bounded universal quantification of a set in P. Thus, the set $most(E, p)$ is in Σ_2 by Lemma 8.20. But, by definition, for words y of length $p(n)$, y belongs to $most(E, p)$ if and only if there are more than $2^{p(n)-n}$ words w of length $p(n)$ such that $\langle y, w \rangle \in E$.

Substituting the definition of E, we obtain that:

$x \in A$ if and only if $\exists^{p(|x|)} y (\langle x, y \rangle \in B$ and $y \in most(E, p))$

But now observe that since B is in P, and $most(E,p)$ is in Σ_2, the predicate in parentheses is in Σ_2. Therefore A has been expressed as an existential quantification of a predicate in Σ_2, and by Theorem 8.8, part (c), A is in Σ_2, as was to be shown. $\qquad\qquad\qquad\qquad\qquad\qquad\qquad\qquad\qquad\qquad\qquad\qquad$ $\square$

8.6 Exercises

1. State and prove all the results stated in this chapter for the relativizations of the polynomial time hierarchy.

2. For any $k \geq 0$, define the set k-QBF as the set of true quantified boolean formulas with at most k alternations of the quantifiers, and starting with an existential quantifier. Prove that for each k, k-QBF is Σ_k-complete.

3. Prove that if $PSPACE \neq PH$ then there are sets in $PSPACE - PH$ which are not $PSPACE$-complete.

4. Prove that for each k, assuming $\Sigma_{k+1} - \Sigma_k$ non-empty, there are sets in this class which are not Σ_{k+1}-complete.

5. Prove that for each k, assuming $\Sigma_{k+1} - \Sigma_k$ non-empty, there are sets in this class which are not NP-hard.

6. Prove that the following classes are not recursively presentable unless they are empty.

   (a) $PSPACE - PH$
   (b) For any $k \geq 0$, $\Sigma_{k+1} - \Sigma_k$
   (c) For any $k \geq 0$, $PSPACE - \Sigma_k$

7. A set A is NP-equivalent if and only if $P(\mathrm{SAT}) = P(A)$. Prove that a set is NP-equivalent if and only if it is Δ_2-complete with respect to the polynomial time Turing reducibility.

8. Prove that if NP is not closed under complements, then there are sets in Δ_2 which are neither NP-equivalent, nor in NP, nor in co-NP. Generalize this statement to other levels of the polynomial time hierarchy.

9. Find a different proof, using Theorem 8.9, of the fact that PH is included in $PSPACE$.

10. Following Definition 5.1 for defining nonuniform complexity classes, define the family of classes $\Sigma_k/poly$, $\Pi_k/poly$, $\Delta_k/poly$, and $PH/poly$. Show that:

    (a) $\Sigma_k/poly = \bigcup_{S \text{ sparse}} \Sigma_k(S)$;
    (b) $\Pi_k/poly = \bigcup_{S \text{ sparse}} \Pi_k(S)$;
    (c) $\Delta_k/poly = \bigcup_{S \text{ sparse}} \Delta_k(S)$;
    (d) $PH/poly = \bigcup_{S \text{ sparse}} PH(S)$.

11. Using the definitions of the previous exercise, show that

$$\Sigma_k/poly = \Pi_k/poly \Rightarrow \Sigma_k/poly = PH/poly$$

and that

$$\Sigma_k/poly = \Pi_k/poly \Rightarrow \Sigma_{k+2} = \Pi_{k+2}$$

12. Define the *exponential time hierarchy relative to A* as follows:
    (a) $\Sigma_0^E(A) = \Pi_0^E(A) = \Delta_0^E(A) = P(A)$;
    (b) $\Sigma_{k+1}^E(A) = NEXT(\Sigma_k(A))$ for $k \geq 0$;
    (c) $\Delta_{k+1}^E(A) = DEXT(\Sigma_k(A))$ for $k \geq 0$;
    (d) $\Pi_{k+1}^E(A) = \text{co-}\Sigma_{k+1}^E(A)$ for $k \geq 0$,

    where $DEXT(A)$ and $NEXT(A)$ were defined in Chapter 3. Define also

    $$EH = (\bigcup_{k \geq 0} \Sigma_k^E(A)) \bigcup (\bigcup_{k \geq 0} \Pi_k^E(A))$$

    The *exponential time hierarchy* is obtained when $A = \emptyset$. Prove that for every k, a set C is in $\Sigma_k^E(A)$ if and only if for some $B \in P(A)$ and for some function $e(n) = 2^{c \cdot n}$, $x \in C$ if and only if

    $$\exists^{e(|x|)} y_1 \forall^{e(|x|)} y_2 \ldots Q y_k \langle x, y_1, \ldots, y_k \rangle \in B$$

    where Q is $\exists^{e(|x|)}$ if k is odd, and $\forall^{e(|x|)}$ if k is even.

13. Prove the following:
    (a) $\Sigma_k^E(A) = \text{co-}\Pi_k^E(A)$
    (b) $\Sigma_k^E(A) \cup \Pi_k^E(A) \subseteq \Sigma_{k+1}^E(A) \cap \Pi_{k+1}^E(A)$
    (c) For every $k \geq 1$, $\Sigma_k(A) = \Pi_k(A)$ implies $\Sigma_k^E(A) = \Pi_k^E(A)$.

14. Let EQUIV be the set of all pairs of encodings of equivalent boolean formulas. Show that EQUIV $\in \Pi_1$.

15. Let MINIMAL be the set of all encodings of boolean formulas F for which there is no expression equivalent to F but shorter than F. Use the previous exercise to show that MINIMAL $\in \Pi_2$.

16. Show that the following statements are equivalent.

    (a) The polynomial time hierarchy collapses;
    (b) For every set $A \in PH$, the polynomial time hierarchy relative to A collapses;
    (c) There is a set $A \in PH$ such that the polynomial time hierarchy relative to A collapses;
    (d) There is a set $A \in PH$ such that $P(A) = NP(A)$.

8.7 Bibliographical Remarks

The polynomial time hierarchy was implicitly proposed in Karp (1972). However, it was made formally explicit for the first time in Meyer and Stockmeyer (1973), as a possible way of classifying problems not known to be in *NP*. They provide the complete sets for each level described in Exercise 2. Exercises 14 and 15 are also from this reference.

The main references for the initial study of the polynomial time hierarchy are Stockmeyer (1977) and Wrathall (1977), where the properties we have stated and the alternating quantifiers characterization are proved. These references contain also other characterizations and properties, such as for example Exercise 2. In our presentation we have followed ideas from Schöning (1981).

The relativizations of the polynomial time hierarchy are proposed in Baker, Gill, and Solovay (1975), and the lower levels will be studied in Volume II. The exercises on the application of the Uniform Diagonalization Theorem to the polynomial time hierarchy are from Schöning (1982) and Schöning (1983); Exercise 8 is proposed as an open problem in Garey and Johnson (1978), and the solution presented here is due to Schöning (1983).

The classes $\Sigma_k/poly$ and $PH/poly$ are sometimes called "the advice hierarchy". These classes are defined and studied in Yap (1983). Exercises 10 and 11 are from this reference.

The exponential time hierarchy presented in Exercises 12 and 13 was defined in Simon (1975), and has been studied later by a number of researchers. Exercises 12 and 13 are from Wilson (1980) and Orponen (1983). Other properties of the relativizations of the exponential time hierarchy can be found in Dekhtyar (1976) and Heller (1984). A different version of the exponential time hierarchy (called "strong exponential time hierarchy") can be defined, but Hemachandra (1989) has shown that this alternative definition yields a collapsing hierarchy.

The fact that BPP is included in the second level of the hierarchy is due to Lautemann (1983) and, simultaneously, to Gács and Sipser—announced without proof in Sipser (1983). Our proof follows Lautemann's ideas, in a more developed fashion, and is partially inspired in the proof of a stronger fact in Schöning (1985a). For closely related results see Zachos and Heller (1986). On the other hand, the relationship of the polynomial time hierarchy with the class PP is far less clear, but it has been discovered by Toda (1991) that the whole hierarchy is included in the polynomial time Turing closure of PP.

Savitch's Theorem (Theorem 2.27) implies that $NPSPACE = PSPACE$, and therefore iterating the operator $NPSPACE$ does not yield a proper hierarchy over $PSPACE$. Thus, such a hypothetical "polynomial space hierarchy" collapses. In Volume II we will use alternating Turing machines to construct a hierarchy in this way. For logarithmic space bounds Savitch's Theorem still admits, in principle, a hierarchy, as defined in Ruzzo, Simon, and Tompa (1984), by iterating a suitable definition of the $NLOG$ operator. The definition requires a somehow artificial definition of space bounded oracle machines. Also in principle, this definition may differ from the one given by alternating Turing machines. However, as a consequence of Theorem 2.26(k), these definitions do not reach beyond *NLOG*. See the proof and the bibliographical remarks in the Appendix.

References

Adleman, L. (1978): "Two theorems on random polynomial time". In: *Proc. 19th. IEEE Symp. on Foundations of Computer Science*, 75–83.

Adleman, L., Manders, K. (1977): "Reducibility, randomness, and intractability". In: *Proc. 9th. ACM Symp. on Theory of Computing*, 151–163.

Adleman, L., Ming-Deh Huang (1987): "Recognizing primes in random polynomial time". In: *Proc. 19th. ACM Symp. on Theory of Computing*, 462–469.

Aho, A.H., Hopcroft, J., Ullman, J. (1974): *The Design and Analysis of Computer Algorithms*. Addison-Wesley, Reading, Mass.

Allender, E.W. (1985): *Invertible Functions*. Ph. D. Dissertation, School of Information and Computer Science, Georgia Institute of Technology.

Ambos-Spies, K. (1986): "Inhomogeneities in the polynomial-time degrees: the degrees of super sparse sets". *Information Processing Letters* **22**, 113–117.

Ambos-Spies, K. (1987): "A note on complete problems for complexity classes". *Information Processing Letters* **23**, 227–230.

Bach, E., Miller, G., Shallit, J. (1984): "Sums of divisors, perfect numbers, and factoring". In: *Proc. 16th. ACM Symp. on Theory of Computing*, 183–190.

Baker, T., Gill, J., Solovay, R. (1975): "Relativizations of the $P =? NP$ question". *SIAM Journal on Computing* **4**, 431–442.

Balcázar, J.L., Book, R.V. (1986): "Sets with small generalized Kolmogorov complexity". *Acta Informatica* **23**, 679–688.

Balcázar, J.L., Book, R.V., Schöning, U. (1986): "The polynomial-time hierarchy and sparse oracles". *Journal ACM* **33**, 603–617.

Balcázar, J.L., Diaz, J. (1982): "A note on a theorem by Ladner". *Information Processing Letters* **15**, 84–86.

Balcázar, J.L., Diaz, J., Gabarró, J. (1985): "Uniform characterizations of nonuniform complexity measures". *Information and Control* **67**, 53–69.

Balcázar, J.L., Diaz, J., Gabarró, J. (1987): "On characterizations of the class PSPACE/poly". *Theoretical Computer Science* **52**, 1–17.

Balcázar, J.L., Gabarró, J. (1986): "Some comments about notations of orders of magnitude". *Bulletin EATCS* **30**, 34–42.

Balcazár, J.L., Gabarró, J. (1989): "Nonuniform complexity classes specified by lower and upper bounds". *Informatique Theorique et Applications* **23**, 177–194.

Beigel, R., Reingold, N., Spielman, D. (1994): "PP is closed under intersection". To appear in *Journal of Computer and System Sciences*.

Bennett, C., Gill, J. (1981): "Relative to random oracle A, $P^A \neq NP^A \neq co\text{-}NP^A$ with probability 1". *SIAM Journal on Computing* **10**, 96–113.

Blum, N. (1984): "A boolean function requiring $3n$ network size". *Theoretical Computer Science* **28**, 337–345.

Book, R.V. (1972): "On languages accepted in polynomial time". *SIAM Journal on Computing* **1**, 281–287.

Book, R.V. (1974a): "Comparing complexity classes". *Journal of Computer and System Sciences* **9**, 213–229.

Book, R.V. (1974b): "Tally languages and complexity classes". *Information and Control* **26**, 186–193.

Book, R.V. (1976): "Translational lemmas, polynomial time, and $(\log n)^{j}$-space". *Theoretical Computer Science* **1**, 215–226.

Book, R.V., Ko, Ker-I (1988): "On sets truth-table reducible to sparse sets". *SIAM Journal on Computing* **17**, 903–919.

Borodin, A. (1973): "Computational complexity: theory and practice". In: *Currents in the Theory of Computing* (A.V.Aho, ed.), 35–89. Prentice-Hall, Englewood Cliffs, NJ.

Borodin, A. (1977): "On relating time and space to size and depth". *SIAM Journal on Computing* **6**, 733–744.

Borodin, A., Cook, S.A., Dymond, P., Ruzzo, W., Tompa, M. (1989): "Two applications of complementation via inductive counting". *SIAM Journal on Computing* **18**, 559–578.

Buntrock, G., Hemachandra, L., Siefkes, D. (1993): "Using Inductive Counting to Simulate Nondeterministic Computation". *Information and Computation* **102**, 102–117.

Chew, P., Machtey, M. (1981): "A note on structure and looking-back applied to the relative complexity of computable functions". *Journal of Computer and System Sciences* **22**, 53–59.

Church, A. (1933): "A set of postulates for the foundation of logic". *Annals of Mathematics* **25**, 839–864.

Church, A. (1936): "An unsolvable problem of elementary number theory". *The American Journal of Mathematics* **58**, 345–363.

Cobham, A. (1964): "The intrinsic computational difficulty of functions" . In: *Proc. Congress for Logic, Mathematics, and Philosophy of Science*, 24–30. North-Holland, Amsterdam.

Cook, S. (1971): "The complexity of theorem proving procedures". In: *Proc. 3rd. ACM Symp. on Theory of Computing*, 151–158.

Cook, S. (1973): "An observation of time-storage trade-off". In: *Proc. 5th. Annual ACM Symposium on the Theory of Computing*, 29–33.

Davis, M.D., Weyuker, E.J. (1983): *Computability, Complexity, and Languages.* Academic Press, New York.

de Leeuw, K., Moore, E.F., Shannon, C.E., Shapiro, N. (1956): "Computability by probabilistic machines". In: *Automata Studies* (C.E. Shannon, ed.), 183–198. Annals of Mathematical Studies, 34. American Mathematical Society, Rhode Island.

Dekhtyar, M. (1976): "On the relativization of deterministic and nondeterministic complexity classes". In: *Mathematical Foundations of Computer Science*

188   References

(A. Mazurkiewicz, ed.), 255–259. Springer-Verlag Lecture Notes in Computer Science 45.

Edmonds, J. (1965): "Paths, trees, and flowers". *Canada Journal of Mathematics* **17**, 449–467.

Fortnow, L., Reingold, N. (1991): "PP is closed under truth-table reductions". In: *IEEE Structure in Complexity Theory 6th Annual Conference*, 13–15.

Gabarró, J. (1983a): *Funciones de Complejidad y su Relación con las Familias Abstractas de Lenguajes*. Doctoral Thesis, Facultat d'Informatica de Barcelona.

Gabarró, J. (1983b): "Initial index: a new complexity function for languages". In: *10th. Int. Coll. on Automata, Languages, and Programming* (J. Diaz, ed.), 226–236. Springer-Verlag Lecture Notes in Computer Science, 154.

Garey, M., Johnson, D. (1978): *Computers and Intractability: A Guide to the Theory of NP-completeness*. Freeman, San Francisco.

Gill, J. (1972): *Probabilistic Turing Machines and Complexity of Computations*. Ph. D. Dissertation, U.C. Berkeley.

Gill, J. (1977): "Computational complexity of probabilistic Turing machines". *SIAM Journal on Computing* **6**, 675–695.

Gödel, K. (1931): "On formally undecidable propositions of Principia Mathematica and related systems". *Monatshefte fur Math. und Physik* **38**, 173–198.

Harper, L.H., Savage, J.E. (1972): "The complexity of the marriage problem". *Advances in Mathematics* **9**, 299–312.

Harrison, M.A. (1978): *Introduction to Formal Languages*. Addison-Wesley, Reading, Mass.

Hartmanis, J. (1983): "On sparse sets in $NP - P$". *Information Processing Letters* **16**, 55–60.

Hartmanis, J., Hunt, H.B. (1974): "The LBA problem and its importance in the theory of computing". In: *SIAM-AMS Proceedings*, vol. 7, 1–26.

Hartmanis, J., Stearns, R.E. (1965): "On the computational complexity of algorithms". *Trans. American Mathematical Society* **117**, 285–306.

Hartmanis, J., Yesha, Y. (1984): "Computation times of NP sets of different densities". *Theoretical Computer Science* **34**, 17–32.

Heller, H. (1984): "On relativized polynomial and exponential computations". *SIAM Journal on Computing* **13**, 717–725.

Hemachandra, L.A. (1989): "The strong exponential hierarchy collapses". *Journal of Computer and System Sciences* **39**, 299–322.

Hennie, F.C., Stearns R.E. (1966): "Two-tape simulation of multitape Turing machines". *Journal ACM* **13**, 533–546.

Hinman, P., Zachos, E. (1984): "Probabilistic machines, oracles, and quantifiers". In: *Proc. Recursion Theoretic Week, Oberwolfach*, 159–192. Springer-Verlag Lecture Notes in Mathematics 1141.

Homer, S. (1981): "Some properties of the lattice of NP sets". In: *Workshop on Recursion Theoretic Aspects of Computer Science*, 18–22. Purdue University.

Hopcroft, J., Ullman, J. (1979): *Introduction to Automata Theory, Languages, and Computation*. Addison-Wesley, Reading, Mass.

Hromkovič, J. (1988): "Two independent solutions of the 23 years old open problem in one year or *NPSPACE* is closed under complementation by two authors". *Bulletin of the EATCS* **34**, 310–312.

Immerman, N. (1988): "Nondeterministic space is closed under complementation". *SIAM Journal on Computing* **17**, 935–938.

Jenner, B., Kirsig, B., Lange, K.-J. (1989): "The logarithmic alternation hierarchy collapses: $A\Sigma_2^{\mathcal{L}} = A\Pi_2^{\mathcal{L}}$". *Information and Computation* **80**, 269–288.

Jockusch, C. (1968): "Semirecursive sets and positive reducibilities". *Trans. American Mathematical Society* **131**, 420–436.

Jones, N.D. (1975): "Space-bounded reducibility among combinatorial problems". *Journal of Computer and System Science* **11**, 68–85.

Jones, N.D., Laaser, W.T. (1976): "Complete problems for deterministic polynomial time". *Theoretical Computer Science* **3**, 105–118.

Jones, N.D., Lien, E., Laaser W.T. (1976): "New problems complete for nondeterministic log space". *Math. Systems Theory* **10**, 1–17.

Kannan, R. (1982): "Circuit-size lower bounds and nonreducibility to sparse sets". *Information and Control* **55**, 40–56.

Karp, R.M. (1972): "Reducibility among combinatorial problems". In: *Complexity of Computer Computations* (R. Miller and J. Thatcher, eds.), 85–104. Plenum Press, New York.

Karp, R.M., Lipton, R.J. (1980): "Some connections between nonuniform and uniform complexity classes". In: *Proc. 12th. ACM Symp. on Theory of Computing*, 302–309.

Kleene, S. (1936): "General recursive functions of natural numbers". *Mathematische Annalen* **112**, 727–742.

Kleene, S. (1952): *Introduction to Metamathematics*. D. Van Nostrand, Princeton, NJ.

Knuth, D. (1976): "Big omicron and big omega and big theta". *SIGACT News* **8**, 18–24.

Ko, Ker-I (1982): "Some observations on the probabilistic algorithms and *NP*-hard problems" *Information Processing Letters*, **14**, 39–43.

Ko, Ker-I (1983): "On self-reducibility and weak P-selectivity". *Journal of Computer and System Sciences* **26**, 209–221.

Ko, Ker-I (1985): "Continuous optimization problems and a polynomial hierarchy of real functions". *Journal of Complexity* **1**, 210–231.

Ko, Ker-I (1987): "On helping by robust oracle machines". *Theoretical Computer Science* **52**, 15–36.

Kobayashi, K. (1985): "On proving time constructibility of functions". *Theoretical Computer Science* **35**, 215–225.

Köbler, J., Schöning, U., Wagner, K. (1987): "The difference and truth-table hierarchies for NP". *Informatique Théorique et Applications* **21**, 419–435.

Kranakis, E. (1986): *Primality and Cryptography*. Wiley-Teubner Series in Computer Science, Stuttgart.

Ladner, R. (1975a): "On the structure of polynomial-time reducibility". *Journal ACM* **22**, 155–171.

Ladner, R. (1975b): "The circuit value problem is log space complete for P". *SIGACT News* **7**, 18–20.

Ladner, R., Lynch N. (1976): "Relativization of questions about log space computability". *Math. System Theory* **10**, 19–32.

Ladner, R., Lynch, N., Selman A., (1975): "A comparison of polynomial-time reducibilities". *Theoretical Computer Science* **1**, 103–123.

Landweber, L., Lipton, R., Robertson, E. (1981): "On the structure of sets in NP and other complexity classes". *Theoretical Computer Science* **15**, 181–200.

Lautemann, C. (1982): "On two computational models for probabilistic algorithms". Preprint.

Lautemann, C. (1983): "*BPP* and the polynomial hierarchy". *Information Processing Letters* **17**, 215–217.

Lewis, H.R., Papadimitriou, C.H. (1981): *Elements of the Theory of Computation*. Prentice-Hall, Englewood Cliffs, NJ.

Lewis, P.M., Stearns, R.E., Hartmanis, J. (1965): "Memory bounds for recognition of context-free and context sensitive languages". In: *Proc. 6th. Annual IEEE Symp. on Switching Circuit Theory and Logical Design*, 191–202.

Long, T. (1982): "Strong nondeterministic polynomial-time reducibilities". *Theoretical Computer Science* **21**, 1–25.

Lupanov, O.B. (1958): "A method of circuit synthesis". *Izvestia V.U.Z. Radiofizika* **1**, 120–140.

Mahaney, S.R. (1982): "Sparse complete sets for NP: solution of a conjecture by Berman and Hartmanis". *Journal of Computer and System Sciences* **25**, 130–143.

McCulloch, W.S., Pitts, E. (1943): "A logical calculus of the ideas immanent in nervous activity". *Bulletin Mathematical Biophysics* **5**, 115–133.

Meyer, A., Paterson, M. (1979): "With what frequency are apparently intractable problems difficult?" *Technical Report*, 126, MIT.

Meyer, A., Stockmeyer, L. (1973): "The equivalence problem for regular expressions with squaring requires exponential time". In: *Proc. 13th. IEEE Symp. on Switching and Automata Theory*, 125–129.

Miller, G.L. (1976): "Riemann's hypothesis and tests for primality". *Journal of Computer and System Sciences* **13**, 300–317.

Orponen, P. (1983): "Complexity classes of alternating machines with oracles". In: *10th. Int. Coll. on Automata, Languages, and Programming* (J. Diaz, ed.), 573–584. Springer-Verlag Lecture Notes in Computer Science 154.

Papadimitriou, C.H., Yannakakis, M. (1982): "The complexity of facets (and some facets of complexity)". In: *Proc. 14th. ACM Symp. on Theory of Computing*, 255–260.

Papadimitriou, C.H., Zachos, E. (1985): "Two remarks on the power of counting". Preprint.

Paul, W. (1978): *Komplexitätstheorie*. Teubner, Stuttgart.

Pippenger, N. (1979): "On simultaneous resource bounds". In: *Proc. 20th. IEEE Symp. on Foundations of Computer Science*, 307–311.

Pippenger, N., Fischer, M.J. (1979): "Relations among complexity measures". *Journal of the ACM* **26**, 361–381.

Plaisted, D.A. (1977): "New *NP*-hard and *NP*-complete polynomial and integer divisibility problems". In: *Proc. 18th. IEEE Symp. on Foundations of Computer Science*, 241–253.

Post, E. (1936): "Finite combinatory process". *Journal of Symbolic Logic* **1**, 103–105.

Pratt, V. (1975): "Every prime has a succint certificate". *SIAM Journal on Computing* **4**, 214–220.

Rabin, M.O. (1976): "Probabilistic algorithms". In: *Algorithms and Complexity: New Directions and Results* (J. Traub, ed.), 21–39. Academic Press, London.

Rabin, M.O., Scott, D. (1959): "Finite automata and their decision problems". *IBM Journal of Research and Development* **3**, 114–125.

Rackoff, C. (1982): "Relativized questions involving probabilistic algorithms". *Journal ACM* **29**, 261–268.

Regan, K. (1983): "On diagonalization methods and the structure of language classes". In: *Fundamentals of Computation Theory* (M. Karpinski, ed.), 368–380. Springer-Verlag Lecture Notes in Computer Science 158.

Reif, J. (1985): "Probabilistic algorithms in group theory". In: *Fundamentals of Computation Theory* (L. Budach, ed.), 341–350. Springer-Verlag Lecture Notes in Computer Science 199

Ritchie, R.W. (1963): "Classes of predictably computable functions". *Trans. American Mathematical Society* **106**, 139–173.

Riordan, J., Shannon, C.E. (1942): "The number of two-terminal series-parallel networks". *Journal of Mathematics and Physics* **21**, 83–93.

Rogers, H. (1967): *Theory of Recursive Functions and Effective Computability.* McGraw-Hill, New York.

Russo, D. (1985): *Structural Properties of Complexity Classes.* Ph. D. Dissertation, U.C. Santa Barbara.

Ruzzo, W., Simon, J., Tompa, M. (1984): "Space-bounded hierarchies and probabilistic computation". *Journal of Computer and System Science* **28**, 216–230.

Santos, E. (1969): "Probabilistic Turing machines and computability". *Proc. American Mathematical Society* **22**, 704–710.

Santos, E. (1971): "Computability by probabilistic Turing machines". *Trans. American Mathematical Society* **159**, 165–184.

Savage, J.E. (1972): "Computational work and time of finite machines". *Journal ACM* **19**, 660–674.

Savage, J.E. (1976): *The Complexity of Computing.* J. Wiley, New York.

Savitch, W.J. (1970): "Relationships between nondeterministic and deterministic tape complexities". *Journal of Computer and System Sciences* **4**, 177–192.

Schmidt, D. (1985): "The recursion-theoretic structure of complexity classes". *Theoretical Computer Science* **38**, 143–156.

Schnorr, C.P. (1976a): "Optimal algorithms for self-reducible problems". In: *3rd. Int. Coll. on Automata, Languages, and Programming* (R. Milner, ed.), 322–337. Edinburgh University Press.

Schnorr, C.P. (1976b): "The network complexity and the Turing machine complexity of finite functions". *Acta Informatica* **7**, 95–107.

Schöning, U. (1981): "A note on complete sets for the polynomial time hierarchy". *Sigact News* **13**, 30–34.

Schöning, U. (1982): "A uniform approach to obtain diagonal sets in complexity classes". *Theoretical Computer Science* **18**, 95–103.

Schöning, U. (1983): "On the structure of Δ_2^P". *Information Processing Letters* **16**, 209–211.

Schöning, U. (1984a): "Minimal pairs for P". *Theoretical Computer Science* **31**, 41–48.

Schöning, U. (1984b): "On small generators". *Theoretical Computer Science* **34**, 337–341.

Schöning, U. (1985a): *Complexity and Structure*. Springer-Verlag Lecture Notes in Computer Science 211.

Schöning, U. (1985b): "Robust algorithms: a different approach to oracles". *Theoretical Computer Science* **40**, 57–66.

Schöning, U., Wagner, K.W. (1988): "Collapsing oracle hierarchies, census functions, and logarithmically many queries". In: *Proc. 5th Symposium on Theoretical Aspects of Computer Science* (R. Cori, M. Wirsing, eds.), 91–98. Lecture Notes in Computer Science 294, Springer-Verlag.

Selman, A. (1978): "Polynomial time enumeration reducibility". *SIAM Journal on Computing* **7**, 440–47.

Selman, A. (1979): "P-selective sets, tally languages and the behavior of polynomial time reducibilities on NP". *Mathematical Systems Theory* **13**, 55–65.

Selman, A. (1982a): "Reductions on NP and p-selective sets". *Theoretical Computer Science* **19**, 287–304.

Selman, A. (1982b): "Analogues of semi-recursive sets and effective reducibilities to the study of NP complexity". *Information and Control* **52**, 36–51.

Shannon, C.E. (1938): "A symbolic analysis of relay and switching circuits". *Trans. AIEE* **57**, 713–723.

Shannon, C.E. (1949): "The synthesis of two-terminal switching circuits". *Bell Systems Technical Journal* **28**, 59–98.

Simon, J. (1975): *On Some Central Problems in Computational Complexity*. Ph. D. Dissertation, Cornell University.

Sipser, M. (1980): "Halting space-bounded computations". *Theoretical Computer Science* **10**, 335–338.

Sipser, M. (1983): "A complexity-theoretic approach to randomness". In: *Proc. 15th. ACM Symp. on Theory of Computing*, 330–335.

Slot, C.F., van Emde Boas, P. (1985): "On tape versus core; an application of space efficient perfect hash functions to the invariance of space". In: *Proc. 17th. ACM Symp. on Theory of Computing*, 391–400.

Solovay, R., Strassen, V. (1977): "A fast Monte-Carlo test for primality". *SIAM Journal on Computing* **6**, 84–85.

Solovay, R., Strassen, V. (1978): "Erratum on 'A fast Monte-Carlo test for primality' ". *SIAM Journal on Computing* **7**, 118.

Stearns, R.E., Hartmanis, J., Lewis, P.M. (1965): "Hierarchies of memory-limited computations". In: *Proc. 6th Annual IEEE Symposium on Switching Circuit Theory and Logical Design*, 179–190.

Stockmeyer, L. (1977): "The polynomial-time hierarchy". *Theoretical Computer Science* **3**, 1–22.

Stockmeyer, L., Meyer, A. (1973): "Word problems requiring exponential time". In: *Proc. 5th. ACM Symp. on Theory of Computing*, 1–9.

Szelepcsényi, R. (1988): "The method of forced enumeration for nondeterministic automata". *Acta Informatica* **26**, 279–284.

Toda, S. (1987): "Σ_2-SPACE is closed under complement". *Journal of Computer and System Sciences* **35**, 145–152.

Toda, S. (1991): "PP is as hard as the polynomial-time hierarchy". *SIAM Journal on Computing* **20**, 865–877.

Trakhtenbrot, B.A. (1984): "A survey of Russian approaches to *perebor* (brute-force search) algorithms". *Annals of the History of Computing* **6**, 384–400.

Turing, A. (1936): "On computable numbers with an application to the 'Entschei-dungsproblem' ". *Proc. London Mathematical Society* **2**, 230–265.

Turing, A. (1937): "Rectification to 'On computable numbers...' ". *Proc. London Mathematical Society* **4**, 544–546.

Uspenski, V.A. (1979): *The Post Machine.* MIR, Moscow.

Valiant, L.G. (1976): "The relative complexity of checking and evaluating". *Information Processing Letters* **5**, 20–23.

Valiant, L.G. (1979): "The complexity of computing the permanent". *Theoretical Computer Science* **8**, 189–201.

Vitányi, P.M.B., Meertens, L.G.L.T. (1984): "Big omega versus the wild functions". *Bulletin EATCS* **22**, 14–19.

Wagner, K., Wechsung, G. (1986): *Computational Complexity.* Reidel, Dordrecht.

Welsh, D. (1983): "Randomized algorithms". *Discrete Applied Mathematics* **5**, 133–145.

Wilson, C. (1980): *Relativization, Reducibilities, and the Exponential Time Hierarchy.* Doctoral Dissertation, Univ. Toronto.

Wrathall, C. (1977): "Complete sets and the polynomial time hierarchy". *Theoretical Computer Science* **3**, 23–33.

Yamada, H. (1962): "Real-time computation and recursive functions not real-time computable". *IEEE Trans. on Electronic Computers* **11**, 753–760.

Yap, C.P. (1983): "Some consequences of non-uniform conditions on uniform classes". *Theoretical Computer Science* **27**, 287–300.

Zachos, E. (1982): "Robustness of probabilistic computational complexity classes under definitional perturbations". *Information and Control* **54**, 143–154.

Zachos, E., Heller, H. (1986): "A decisive characterization of *BPP*". *Information and Control* **69**, 125–135.

Appendix
Complementation via Inductive Counting

1 Nondeterministic Space is Closed Under Complement

The closure under complementation of nondeterministic complexity classes is still a research problem. Several computational notions, in particular those regarding the power of alternating quantifiers (as in the polynomial time hierarchy) or of resource-bounded alternation-bounded alternating machines, will not be well understood without solving this problem: for instance, if NP is closed under complementation then the polynomial time hierarchy collapses to Σ_1.

We present in this appendix a technique due, independently, to N. Immerman and R. Szelepcsényi to show that nondeterministic space classes are closed under complement. The technique consists of nondeterministically counting, in an inductive manner, the number of accessible configurations and the number of accepting accessible configurations. The fundamental property of the technique is that knowing the number N of accessible configurations allows one to check, when it is the case, that no accepting accessible configuration exists, by nondeterministically finding N non-accepting accessible configurations.

The main tool is presented in the next lemma, which shows how to decide whether a given configuration is accessible in $d + 1$ steps provided that the number of configurations accessible in d steps is known. The key property of the nondeterministic machine that does so is that on the appropriate inputs it is a strong nondeterministic machine (see Exercise 24 in Chapter 4). We say that a nondeterministic machine with accepting and rejecting final states is *strong on input x* if either some computation accepts and no computation rejects, or some computation rejects and no computation accepts. Thus, it does not yield contradictory answers. The interest of strong machines is that they are always able to provide a correct answer, if the nondeterminism works properly, and if it fails they can detect it and abort the computation. This fact is used in the proof.

Our notion of *configuration* in this appendix is that presented in Definition 1.44, in which the contents of the input tape is omitted; the context will make clear at each moment what the input is. Let M be a nondeterministic space-bounded machine with space bound $s(n)$. If it is at least logarithmic, then $O(s(n))$ space suffices for describing each configuration; thus we identify each configuration with a word of length $O(s(n))$ which encodes it. Cycling over all the words of this length, e.g. in lexicographical order (by repeatedly adding 1 in binary), allows one to cycle over all the possible configurations of M on inputs of length n. Cycling in this manner will be very useful in the algorithms in the proof.

Denote by I_0 the (encoding of the) initial configuration of M, which is independent of the input. For an input x, denote by $Rch_{M,x}(t)$ the number of configurations reachable by computations on input x in at most t steps starting at I_0. Again, when M and x are easily identifiable from the context, the subscript will be omitted. Finally, let $t(n)$ be a time bound for M obtained as in Theorem 2.26(g).

We now describe a nondeterministic decision procedure for reachability of configurations. The major property is that when the input provides the correct number $Rch_{M,x}(t)$ of configurations, then the procedure is strong: some computation does not abort, and all nonaborting computations give a correct answer.

Lemma 1 *Given M as before, there is a nondeterministic machine M_0 working in space $s(n)$, which on each input $\langle x, n, t, u \rangle$, where $n = Rch(t)$, is strong and accepts if and only if u encodes a configuration reachable in at most $t + 1$ steps for I_0.*

Proof. The machine is presented in Figure 1. Assume that some computation reaches u within $t + 1$ steps, and let v be a predecessor of u in this computation. Then the nondeterministic choices that hit on the the right way of reaching v in at most t steps will accept; any other nondeterministic choice will fail to count v as accessible, and therefore will end up with $m < n$, thus aborting. No computation rejects.

Conversely, if u is not reachable then no computation will accept; moreover, some nondeterministic choices will find all the right ways of reaching each configuration v reachable in at most t steps, and thus the machine will reach $m = n$ and reject. Thus if $n = Rch(t)$ then M_0 is strong. $\quad\square$

The machine works properly for every $t \geq 0$. In particular for $t = 0$ the only configuration reachable in 0 steps is I_0, n is expected to have value 1, and M_0 will answer YES if and only if u is reachable in one step from I_0.

In the main proof, this lemma is used for two purposes: (1) to inductively count the number of reachable configurations, by computing $Rch(t+1)$ from each $Rch(t)$ up to the time bound $t(n)$; and (2) to check whether any accepting configuration is reachable within the time bound. The important point

---

M_0:
input x, n, t, u
$m := 0$
for each word v coding a configuration of M do
    nondeterministically simulate M on x during at most t steps,
        checking whether v is reached
    if so, then
        $m := m + 1$
        if M on input x goes in one step from v to u then accept
    end for
if $m = n$ then reject
else abort by halting in a nonfinal ("?") state
end

Figure 1   A machine to test reachability

---

is that the "negative" information provided by M_0 is complete, in the sense that its status as a strong nondeterministic machine for the proper inputs guaranteees that the answer is NO if and only if the configuration u received as input is NOT accessible from the starting configuration.

We now present the machine that uses M_0 to inductively count the number of accessible configurations $Rch(t(|x|))$. It computes it nondeterministically; this means that there are always some nondeterministic paths that correctly compute the number, and that all the others detect that they fail to compute the correct number and halt without answer.

Lemma 2 *On the same hypothesis as the previous lemma, there is a machine which on input x nondeterministically computes the number of accessible configurations $Rch(t(|x|))$.*

Proof. This machine repeatedly calls M_0, but is deterministic otherwise. It is presented in Figure 2. The comments indicate the assertions required to verify that it meets the statement of the lemma.

The assertion $n = Rch(t)$ guarantees that all the calls to M_0 are correct, and therefore it either aborts (and then M_1 aborts as well) or answers correctly. Thus at the end of the external loop t the value of m has correctly counted how many configurations were accessible in at most $t + 1$ steps, and the last assignment updates n so as to maintain the invariant. Thus, if M_1 does not abort then it computes $Rch(t(|x|))$ as required.   □

The information computed by M_1 allows us to use M_0 again to check the reachability of accepting configurations, with the certainty that M rejects if

```
input x
n := 1
t := 0
for each t from 1 to t(|x|) do
comment: assertion n = Rch(t) is invariant for the loop
      for each u encoding a configuration do
             if u is reachable from I_0 in at most t + 1 steps then m := m + 1
comment: this is tested by calling M_0 on ⟨x, n, t, u⟩
      end internal loop
comment: reestablish invariant by updating n
      n := m
comment: ready for the next t
end
```

Figure 2   Computing the number of accessible configurations

and only if some sequence of nondeterministic choices leads M_0 to reject all of them, and no sequence of nondeterministic choices leads M_0 to accept any of them. This can be tested in nondeterministic space $s(n)$.

Theorem 3   *If $s(n) \geq \log n$ and s is space constructible then NSPACE(s) is closed under complementation.*

Proof.   Let M accept L in nondeterministic space s and in time t. Consider a machine that on input x calls M_1 on x to compute $n = Rch_{M,x}(t|x|)$, and then for each accepting configuration u of M calls M_0 on input $\langle x, n, t(|x|), u \rangle$ to check whether any of them is accessible, accepting if no accessible accepting configuration exists. This machine accepts $\overline{L}$.   □

A complete machine for the complement of L is presented in Figure 3; it combines the previously presented machines. There, the testing for accessibility of accepting configurations is done simultaneously with the inductive counting.

The algorithm uses several counters. These can be implemented by laying out $s(n)$ cells in any of the working tapes of the machine. This can be done since $s(n)$ is space constructible. K will count the number of steps we are dealing with at any given moment of the algorithm; D will keep track of the number of reachable configurations at each value of K. As usual, the fact that there exists a path of K or less steps, from configuration I_0 to configuration I, will be denoted by $I_0 \vdash^{\leq k} I$. Counters C_1 and C_2 will contain at some moment the number of configurations reachable in at most $K - 1$ and K steps respectively.

comment: Machine M and space bound s are given
input x
create counters D, C_1, C_2, K, and C_3 of size $s(|x|)$
compute initial ID I_0
deterministically compute the number I_1 of configurations
  reachable in one step from I_0
$C_2 := 0$
for each I do
  if deterministically $I_0 \vdash^1 I$ then $C_2 := C_2 + 1$
$D := C$
comment: K counts up to the maximum number of steps
$K := 2$
while $K < 2^{s(n)}$
  $C_2 := 0$
comment: C_2 counts up to $Rch(K)$
  for each configuration I_2 do
comment: check whether I_2 is reachable in k steps
  $C_1 := 0$
comment: C_1 counts up to $Rch(K-1)$
   for each configuration I_1 do
comment: check whether I_1 is reachable in $k-1$ steps and precedes I_2
    guess path $I_0 \vdash^{\le k} I_1$
    if there exists a correct path, then $C_1 := C_1 + 1$
    if $I_1 \vdash^1 I_2$, then
     if I_2 is accepting, REJECT and STOP
     else
      $C_2 := C_2 + 1$
      exit the inner loop
   end for loop on I_1
comment: check that the nondeterminism worked correctly
   if $C_1 \neq D$ then STOP
  end for loop on I_2
  if $D = C_2$ then ACCEPT
comment: because no new configurations were reached
  else
   $K := K + 1$
   $D := C_2$
  end while
end

Figure 3   Nondeterministic machine accepting $\overline{L(M)}$

2 Bibliographical Remarks

The closure of nondeterministic space classes under complementation via inductive counting was proved independently and almost simultaneously by Immerman (1988) and Szelepcsényi (1988); both results were distributed as internal research reports in 1987. See also Hromkovič (1988). The result followed several previous results proving the collapse of space-bounded hierarchies, shown in Jenner, Kirsig, and Lange (1989), Toda (1987), and Schöning and Wagner (1988); technical reports or conference versions of all these existed by 1987 or earlier. Similar techniques for counting words in certain sets (e.g. sparse sets) were already known; see e.g. Mahaney (1982). Related results are the closure of $NLOG$ under NC reductions—attributed to S. Buss in Immerman (1988)—and the closure under complements of the class LOG(CFL) of the sets $\leq_m^{log}$-reducible to context-free languages—attributed to M. Tompa in Immerman (1988).

Also in Immerman (1988), some consequences are stated regarding closure properties of classes of sets definable in certain first-order languages. The technique of inductive counting has been used also to show properties of classes corresponding to symmetric logarithmic space, probabilistic logarithmic space, and circuits with semi-unbounded fan-in. See Borodin, Cook, Dymond, Ruzzo, and Tompa (1989). Other applications of the technique are given in Buntrock, Hemachandra, and Siefkes (1993).

Author Index

Symbol Index

Index

Former volumes appeared as
EATCS Monographs on Theoretical Computer Science

Vol. 1: K. Mehlhorn
Data Structures and Algorithms 1:
Sorting and Searching

Vol. 5: W. Kuich, A. Salomaa
Semirings, Automata, Languages

Vol. 6: H. Ehrig, B. Mahr
Fundamentals of Algebraic Specification 1
Equations and Initial Semantics

Vol. 7: F. Gécseg
Products of Automata

Vol. 8: F. Kröger
Temporal Logic of Programs

Vol. 9: K. Weihrauch
Computability

Vol. 10: H. Edelsbrunner
Algorithms in Combinatorial Geometry

Vol. 12: J. Berstel, C. Reutenauer
Rational Series and Their Languages

Vol. 13: E. Best, C. Fernández C.
Nonsequential Processes
A Petri Net View

Vol. 14: M. Jantzen
Confluent String Rewriting

Vol. 15: S. Sippu, E. Soisalon-Soininen
Parsing Theory
Volume I: Languages and Parsing

Vol. 16: P. Padawitz
Computing in Horn Clause Theories

Vol. 17: J. Paredaens, P. De Bra, M. Gyssens,
D. Van Gucht
The Structure of the Relational Database
Model

Vol. 18: J. Dassow, G. Páun
Regulated Rewriting in Formal Language
Theory

Vol. 19: M. Tofte
Compiler Generators
What they can do, what they might do,
and what they will probably never do

Vol 20: S. Sippu, E. Soisalon-Soininen
Parsing Theory
Volume II: LR(k) and LL(k) Parsing

Vol 21: H. Ehrig, B. Mahr
Fundamentals of Algebraic Specification 2
Module Specifications and Constraints

Vol. 22: J. L. Balcázar, J. Díaz, J. Gabarró
Structural Complexity II

Vol. 23: A. Salomaa
Public-Key Cryptography

Vol. 24: T. Gergely, L. Úry
First-Order Programming Theories

Vol. 25: W. Wechler
Universal Algebra
for Computer Scientists

R. Janicki, P. E . Lauer
Specification and Analysis of Concurrent
Systems
The COSY Approach
O. Watanabe (Ed.)
Kolmogorov Complexity and Computatio-
nal Complexity

K. Jensen
Coloured Petri Nets
Basic Concepts, Analysis Methods and
Practical Use, Vol. 1

G. Schmidt, Th. Ströhlein
Relations and Graphs
Discrete Mathematics for Computer Scientists

S. L. Bloom, Z. Ésik
Iteration Theories
The Equational Logic of Iterative Processes

Monographs in Theoretical Computer Science – An EATCS Series

C. Calude
Information and Randomness
An Algorithmic Perspective

A. Nait Abdallah
The Logic of Partial Information

K. Jensen
Coloured Petri Nets
Basic Concepts, Analysis Methods and
Practical Use, Vol. 2

Texts in Theoretical Computer Science – An EATCS Series

J. L. Balcázar, J. Díaz, J. Gabarró
Structural Complexity I
2nd ed. (see also below, Vol. 22)

M. Garzon
**Massively Parallel Models of
Computation**
Analysis of Cellular Automata and Neural
Networks